ADVANCED FORMAL VERIFICATION

Advanced Formal Verification

Edited by

Rolf Drechsler
University of Bremen,
Germany

KLUWER ACADEMIC PUBLISHERS
BOSTON / DORDRECHT / LONDON

A C.I.P. Catalogue record for this book is available from the Library of Congress.

ISBN 978-1-4419-5420-6 e-ISBN 978-1-4020-2530-3

Published by Kluwer Academic Publishers,
P.O. Box 17, 3300 AA Dordrecht, The Netherlands.

Sold and distributed in North, Central and South America
by Kluwer Academic Publishers,
101 Philip Drive, Norwell, MA 02061, U.S.A.

In all other countries, sold and distributed
by Kluwer Academic Publishers,
P.O. Box 322, 3300 AH Dordrecht, The Netherlands.

Printed on acid-free paper

Printed in the Netherlands.

Contents

Preface

Modern circuits may contain up to several hundred million transistors. In the meantime it has been observed that verification becomes the major bottleneck in design flows, i.e. up to 80% of the overall design costs are due to verification. This is one of the reasons why recently several methods have been proposed as alternatives to classical simulation. Simulation alone cannot guarantee sufficient coverage of the design resulting in bugs that may remain undetected.

As alternatives formal verification techniques have been proposed. Instead of simulating a design the correctness is proven by formal techniques. There are many different areas where these approaches can be used, like equivalence checking, property checking or symbolic simulation. Meanwhile these methods have been successfully applied in many industrial projects and have become the state-of-the-art technique in several fields. But the deployment of the existing tools in real-world projects also showed the weaknesses and problems of formal verification techniques. This gave motivating impulses for tool developers and researchers.

The book shows latest developments in the verification domain from the user and from the developer perspective. World leading experts describe the underlying methods of today's verification tools and describe various scenarios from industrial practice. In the first part of the book the core techniques of today's formal verification tools, like SAT and BDDs are addressed. In addition, instances known to be difficult, like multipliers, are studied. The second part gives insight in professional tools and the underlying methodology, like property checking and assertion based verification. Finally, to cope with complete system on chip designs also analog components have to be considered.

In this book the state-of-the-art in many important fields of formal verification is described. Besides the description of the most recent research results, open problems and challenging research areas are addressed. By this, the book is intended for CAD developers and researchers in the verification domain, where formal techniques become a core technology to successful circuit and system design. Furthermore, the book is an excellent reference for users of verification tools to get a better understanding of the internal principles and by this to drive the tools to the highest performance. In this context the book is dedicated to all people in industry and academia to keep informed about the most recent developments in the field of formal verification.

Acknowledgment

All contributions in this edited volume have been anonymously reviewed. I would like to express my thanks for the valuable comments of the reviewers and their fast feedback, that allowed a timely publication. Here, I also like to thank all the authors who did a great job in submitting contributions of very high quality. My special thanks go to Görschwin Fey and Daniel Große from my group in Bremen in helping with the preparation of the book. Finally, I would like to thank Cindy Zitter and Mark DeJong from Kluwer Academic Publishers. All this would not have been possible without their steady support.

ROLF DRECHSLER

Contributing Authors

Raik Brinkmann received his Masters Degree in Computer Science from Clausthal Technical University, in 1996. Before joining Infineon, he worked as verification engineer and embedded system designer at Siemens Information and Communication Networks. In 1999 he joined Siemens Corporate Technology to focus his research on formal methods for word-level verification, spending two years at the Infineon Design Center in San Jose, CA. He is also working towards his Ph.D. from University of Kaiserslautern. Currently, he is responsible for verification core technologies in the Infineon CVE formal verification team.

Gianpiero Cabodi received the MS degree in EECS in 1984 and the Ph.D. degree in 1989 from Politecnico di Torino. Since 1989 he has been with the Department of Automation and Computer Engineering of Politecnico di Torino, where he is currently an Associate Professor. He has worked within several EEC funded research projects, and he has been principal investigator of research contracts with DEC, COMPAQ and INTEL. His research interests cover a broad range, within the general framework of CAD for digital systems. He is well known for his scientific contributions within the field of Binary Decision Diagrams applied to Formal Verification. He is also interested in Boolean Satisfiability, Logic and High Level Synthesis, Testing, CAD applications in Parallel and Distributed environments.

Claudionor Nunes Coelho Jr. has a degree in Electrical Engineering (summa cum laude) and a Masters in Computer Science from the Federal University of Minas Gerais. He also holds a Ph.D. in Electrical Engineering and Computer Science from Stanford University. Claudionor was a member of the technical team of several start-up companies, working in an upper management position at Verplex Systems. He is a founder of several start-ups, including RDBIOTEC S.A. and i-Vision. He is also an advisor for FIR Capital Partners. He is currently a professor of Computer Science at the Federal University of Minas Gerais. His interests include validation techniques for complex designs, embedded systems' design and software engineering.

Rolf Drechsler received his diploma and Dr. phil. nat. degree in computer science from the J.W. Goethe-University in Frankfurt am Main, Germany, in 1992 and 1995, respectively. He was with the Institute of Computer Science at the Albert-Ludwigs-University of Freiburg im Breisgau, Germany from 1995 to 2000. He joined the Corporate Technology Department of Siemens AG, Munich in 2000, where he worked as a Senior Engineer in the formal verification group. Since October 2001 he has been with the University of Bremen, Germany, where he is now a full professor for computer architecture. His research interests include verification, logic synthesis, and evolutionary algorithms.

Harry Foster serves as Chairman of the Accellera Formal Verification Technical Committee, which is currently defining the PSL property specification language standard. He is co-author of the Kluwer Academic Publishers book Assertion-Based Design, as well as the Kluwer book Principles of Verifiable RTL Design. Prior to joining Jasper Design Automation, Harry was Verplex Systems' Chief Architect. Harry has researched and developed formal verification tools and methodologies for over 12 years as a Senior Member of the CAD Technical Staff at Hewlett-Packard, and is the original co-creator of the Accellera Open Verification Library (OVL) assertion monitor standard.

Eugene Goldberg received his M.S. degree in theoretical physics from the Belorussian State University in 1983 and his Ph.D. degree in computer science from the Institute of Engineering Cybernetics of the Belorussian Academy of Sciences in 1995. From 1983 to 1995 he worked as a researcher in the laboratory of logic design at the Institute of Engineering Cybernetics. From 1996 to 1997 he was a visiting scholar at the University of California at Berkeley. He joined Cadence Berkeley Labs in November 1997. His main interests are development of efficient algorithms for computationally hard problems with emphasis on CAD applications.

Walter Hartong was born in Dinklage, Germany, on February 10, 1972. He graduated (Dipl.-Ing.) in electrical engineering at the University of Hannover in 1997. He was with the Institute of Microelectronic Circuits and Systems of the University of Hannover since 1997, where he received his Ph.D. degree for his research on approaches to model checking for nonlinear analog systems in 2002. Since 2002 he is application engineer for analog/mixed signal circuit simulation at Cadence Design Systems, Munich, Germany. His research interests include: analog/mixed signal simulation, analog hardware description languages, behavioral modeling, symbolic analysis, and formal verification.

Lars Hedrich was born in Hannover, Germany, on February 19, 1966. He graduated (Dipl.-Ing.) in electrical engineering at the University of Hannover in 1992. Since 1992 he is with the Institute of Microelectronic Circuits and Systems at the Department of Computer Science of the University of Hannover, where he received his Ph.D. degree in electrical engineering for his research on approaches to formal verification of analog circuits in 1997. In 2002 he became a junior professor. His research interests include several areas of analog design automation: symbolic analysis of linear and nonlinear circuits, behavioral modeling, circuit synthesis, and formal verification.

Peer Johannsen received his Masters Degree in Computer Science at Christian-Albrechts-University of Kiel, Germany, in 1997. Afterwards he joined Siemens Corporate Research in Munich, focusing his work on formal techniques for hardware verification. He spent two years of research at the Infineon Design Center in San Jose, CA, working on new methods for property checking of digital circuits. In 2003 he received his Ph.D. in Computer Science from Christian-Albrechts-University of Kiel. Currently he is a project leader in the Infineon CVE formal verification team, responsible for the development of a new static verification tool.

Evgeny Karibaev received the B.S. and Dipl.-Ing. degrees in Electrical Engineering from the Department of Radio Physics at Tomsk State University, Tomsk, Russia, in 2000 and 2001, respectively. He is currently working toward his Ph.D. degree at the Dept. of Electrical and Computer Engineering at the University of Kaiserslautern. His research interests are in the field of formal hardware verification, including equivalence checking and property checking of arithmetic circuits.

Ralf Klausen was born in Hannover, Germany, on May 26, 1971. He graduated (Dipl.-Ing.) in electrical engineering at the University of Hannover in 2000. Since 2000 he is with the Institute of Microelectronic

Systems at the Department of Computer Science of the University of Hannover. He is working towards a Ph.D. degree on approaches to formal verification of analog circuits.

Irina Kufareva graduated from the department of Computer Science of Tomsk State University, Russia, with a Dipl.-Ing. degree in computer science in 1994. She received her Ph.D. in computer science in 2000. She currently works as an assistant professor at the Dept. of Radio Physics at Tomsk State University, Tomsk, Russia. Her current research interests include automata theory and formal verification.

Wolfgang Kunz obtained the Dipl.Ing. degree of electrical engineering from University of Karlsruhe in 1989 and the doctor's degree from University of Hannover in 1992. From 1989 to 1991 he was a graduate student at the ECE Department at the University of Massachusetts at Amherst. From 1993 to 1998 he was with Max Planck Fault-Tolerant Computing Group at the University of Potsdam. From 1998 to 2001 he was a professor at the CS department at the University of Frankfurt. Since 2001 he is with the EE department at the University of Kaiserslautern. Wolfgang Kunz conducts research in the areas of logic and layout synthesis, equivalence checking and ATPG. For his contributions in these areas he has received several awards including the IEEE Transactions on CAD Best Paper Award.

Stefano Quer received the MS degree in EECS in 1991 from Politecnico di Torino and the Ph. D. degree in 1996. In 1994, he was with the EECS Department of the University of California at Berkeley, in 1998, with the *Advanced Technology Group*, at Synopsys Inc., Mountain View, California, and in 1999, with the *Alpha Development Group*, at Compaq, Shrewsbury, Massachussetts. He has been consultant for Compaq Computer Corporation. He is currently Assistant Professor at Dipartimento di Automatica e Informatica of Politecnico di Torino. His research interests include hardware description languages, logic synthesis, formal verification, simulation and testing of digital circuits and systems.

Dominik Stoffel obtained his Diplom-Ingenieur degree from the University of Karlsruhe in 1992 and his Ph.D. from the University of Frankfurt in 1999. From 1994 to 1998 he was with the Max-Planck Fault-Tolerant Computing Group in Potsdam. From 1998 to 2001 he was with the Electronic Design Automation group at the University of Frankfurt,

Germany. Since 2001 he is working as a post-doctoral researcher in the Electronic Design Automation group at the University of Kaiserslautern. His research interests are in the field of logic synthesis and formal hardware verification.

Klaus Winkelmann received a degree in Mathematics and, in 1984, his Ph.D. in computer science at Erlangen University. Currently he is a project manager in the Infineon CVE formal verification team, responsible for innovative applications of property checking technology. Before joining Infineon, he built up and led several R&D teams at Siemens Corporate Research, focusing on formal techniques for the design and verification of embedded software, in particular verification and synthesis of finite automata, synchronous languages, discrete event systems and their applications to industrial control. He also contributed to the application of AI techniques to problems of automation, control and diagnosis. He has acted as reviewer, project manager and as technical director for several European projects, and taught computer science courses in several universities.

Introduction

Rolf Drechsler

With increasing design complexity, verification becomes a more and more important aspect of the design flow. Modern circuits contain up to several million transistors. In the meantime it has been observed that verification becomes the major bottleneck, i.e. up to 80% of the overall design costs are due to verification. This is one of the reasons why recently several methods have been proposed as alternatives to classical simulation, since it cannot guarantee sufficient coverage of the design. E.g. in [2] it has been reported that for the verification of the Pentium IV more than 200 billion cycles have been simulated, but this only corresponds to 2 CPU minutes, if the chip is run with 1 GHz.

Formal verification techniques have gained large attention, since they allow to prove the correctness of a circuit, i.e. they ensure 100% functional correctness. Besides being more reliable, formal verification approaches have also shown to be more cost effective in many cases, since test bench creation - usually a very time consuming and error prone task - becomes superfluous.

In this introduction, we first briefly describe some of the application domains, where formal techniques have successfully been used. We give some links to further literature where the interested reader can get more information. Then, a list of "challenging problems" is given, i.e. a list of topics that need further investigation in the context of formal hardware verification. Finally, the contributions to this book are briefly described.

1. Formal Verification

The main idea of formal hardware verification is to prove the functional correctness of a design instead of simulating some vectors. For the proof process different techniques have been proposed. Most of them

R. Drechsler (ed.), Advanced Formal Verification, xix-xxv.

work in the Boolean domain, like *Binary Decision Diagrams* (BDDs) or SAT solvers.

The typical hardware verification scenarios where formal proof techniques are applied are

Equivalence Checking (EC) and

Property Checking (PC), also called *Model Checking* (MC).

The goal of EC is to ensure the equivalence of two given circuit descriptions. These circuits might be given on different levels of abstraction, i.e. register transfer level or gate level. The main steps of an equivalence checker are as follows (see e.g. [12]):

1. Translate both designs to an internal format.
2. Establish the correspondence between the two designs in a matching phase.
3. Prove equivalence or inequivalence.
4. In case of an inequivalence a counter-example is generated and the debugging phase starts.

Notice that the circuit is considered as purely combinational by modeling the state elements as additional primary inputs and outputs. This modeling may result in counter-examples that are not reachable during normal circuit operation.

In contrast to EC, where two circuits are considered, for PC a single circuit is given and properties are formulated in a dedicated "verification language". It is then formally proven whether these properties hold under all circumstances. While "classical" CTL-based model checking [6] can only be applied to medium sized designs, approaches based on *Bounded Model Checking* (BMC) as discussed in [4] give very good results when used for complete blocks with up to 100k gates.

Nevertheless, all these approaches can run into problems caused by complexity, e.g. if the circuit becomes too large or if the function being represented turns out to be "difficult" for formal methods. The second problem often arises in cases of complex arithmetics, like multipliers.

Motivated by this, hybrid methods have been proposed, like e.g. *symbolic simulation* and *assertion checking*. These methods try to bridge the gap between simulation and correctness proofs. But these techniques also make use of formal proof techniques.

For more information on basics on formal verification techniques the reader is referred to [22].

2. Challenges

Even though formal verification techniques are very successfully applied and have become the state-of-the-art in many design flows, still many problems exist. In this section a list of these problems is given. The list is not complete in the sense that all difficulties are covered, but many important ones are mentioned. This gives a better understanding of current problems in hardware verification, motivates for the following chapters of the book and shows directions for future research.

Complexity: According to Moore's law the complexity of the circuits steadily increases. For this, the underlying data structures are very important. For EC and BMC often dedicated data structures are used. For representation of the state space BDDs have shown to work well, but if the size of the circuit becomes too large the BDDs often suffer from "memory explosion".

Proof technology: While BDDs and SAT are the most popular techniques in hardware verification and have also been applied to many domains, there is still a lot of research going on (see also Chapter 1 and 2). Besides the classical monolithic approaches modern EC tools make use of multi-engine approaches that combine different techniques, like SAT, BDD, term rewriting, ATPG, and random pattern simulation. How to successfully combine these - often orthogonal - approaches is not fully understood today.

Word-level approaches: Even though most proof techniques today work on the bit-level, many studies have shown that significant improvements can be achieved if the proof engine makes use of high-level information or even completely works on a higher level of abstraction. In this context also ILP solvers showed promise (see also Chapter 4).

Matching in EC: As described above, before the proof process starts the correspondence between the circuits has to be established. Here, several techniques exist, like name-based, structural or prover-based, but still for large industrial designs these methods often fail. This results in very time consuming user defined matching.

Reachability of counter-examples: In EC and BMC the generated counter-example might not be reachable in normal circuit operation. This results from the modeling of the circuit, i.e. instead of a FSM only the combinational part is considered. Thus, it has to be checked that the counter-example is "valid" after it has been generated, or the prover has to ensure that it is reachable. Techniques have to be developed how this can be ensured without a complete reachability analysis of the FSM, that is usually not feasible due to complexity reasons.

Arithmetic: Industrial practice has shown that today's proof techniques, like BDD and SAT, have difficulties with arithmetic circuits, like multipliers. Word-level approaches have been proposed as an alternative, but these methods turned out to often be difficult to integrate in fully automatic tools. For this, arithmetic circuits - often occurring in circuit design - are still difficult to handle (see Chapter 4).

System integration: PC works best on the module level, i.e. for blocks with up to 100k gates. But in multi-chip modules many of these blocks are integrated to build a system. Due to complexity the modules cannot be verified as one large block and for this models and approaches are needed.

Hybrid approaches: For complex blocks or on the system level PC might be a very complex task and for this simpler alternatives have been studied, i.e. techniques that are more powerful than classical simulation but need less resources than PC. Techniques, like symbolic simulation or assertion-based verification, in this context also make use of formal verification techniques (see also Chapter 5).

Checker synthesis: The specified properties can also be synthesized and added to the design. In this way, they can also be used for on-line test after the circuit has been fabricated.

Analog/mixed signal: Most EC and PC models assume that the circuit is purely digital, while in modern system-on-chip designs many analog components are integrated. For this, also models and proof mechanisms need to be developed for analog and mixed signal devices (see Chapter 6).

Retiming: For EC retimed circuits are still difficult to handle, since in this case the state matching cannot be performed. Thus, the problem remains sequential and by this becomes far too complex.

Multiple clocks: Many circuits have different clocking domains, while verification tools can often only work with a single clock.

Coverage: To check the completeness of a verification process coverage metrics have to be defined. While typical methods, like state coverage, are much too weak in the context of formal verification, there still does not exist a good measure that is comfortable to use for PC.

Diagnosis: After a fault has been identified by a formal verification tool a counter-example is generated. The next step is to identify the fault location or a reason for the failing proof process. Here, also formal proof techniques can be applied.

Most solutions to these problems are still in a very early stage of development, but these fields have to be addressed to make formal hardware verification successful in industrial applications. To orient the reader, some recent references are provided to give a starting point for further studies: [25, 17, 22, 16, 9, 26, 13, 1, 7, 23, 21, 15, 5, 19, 24, 20, 11, 18, 27, 3, 14, 10, 8]

3. Contributions to this Book

The book consists of six chapters that cover most of the aspects described above. Examples of proof technology are described and the latest developments in this field are presented. But also contributions from industrial practice show the importance of formal verification approaches in today's design flows. Each chapter provides experimental results and for each application domain open problems and directions for future work are outlined.

In Chapter 1, Eugene Goldberg analyses the core problem in formal techniques, i.e. the satisfiability problem. Resolution-based SAT solvers are analyzed and a new way of testing satisfiability is proposed.

Properties of SAT and BDDs are studied in Chapter 2 by Gianpiero Cabodi and Stefano Quer. Based on this analysis, the integration of the two currently most successful proof techniques is discussed.

As mentioned above, formal proof techniques often have difficulties in handling arithmetic circuits. This issue is addressed in Chapter 3 by Dominik Stoffel, Evgeny Karibaev, Irina Kufareva and Wolfgang Kunz, where EC approaches are presented.

New innovative proof techniques that make use of word-level information are described by Raik Brinkmann, Peer Johannsen and Klaus Winkelmann, and an industrial property checking flow is presented in Chapter 4.

In Chapter 5, Claudionor Nunes Coelho Jr. and Harry D. Foster focus on assertion-based verification and in this context introduce a formal property language. The underlying methodology is introduced and implications for the user are addressed.

Finally, an approach to formal verification of analog circuits is proposed in Chapter 6 by Walter Hartong, Ralf Klausen and Lars Hedrich. MC and EC techniques for nonlinear analog systems are discussed.

References

[1] L. Bening and H. Foster. *Principles of Verifiable RTL Design.* Kluwer Academic Publishers, 2001.

[2] B. Bentley. Validating the Intel Pentium 4 microprocessor. In *Design Automation Conf.*, pages 244–248, 2001.

[3] J. Bergeron. *Writing Testbenches: Functional Verification of HDL Models.* Kluwer Academic Publishers, 2003.

[4] A. Biere, A. Cimatti, E.M. Clarke, M. Fujita, and Y. Zhu. Symbolic model checking using SAT procedures instead of BDDs. In *Design Automation Conf.*, pages 317–320, 1999.

[5] R. Brinkmann and R. Drechsler. RTL-datapath verification using integer linear programming. In *ASP Design Automation Conf.*, pages 741–746, 2002.

[6] J.R. Burch, E.M. Clarke, K.L. McMillan, and D.L. Dill. Sequential circuit verification using symbolic model checking. In *Design Automation Conf.*, pages 46–51, 1990.

[7] H. Chockler, O. Kupferman, R. Kurshan, and M. Vardi. A practical approach to coverage in model checking. In *Computer Aided Verification*, volume 2102 of *LNCS*, pages 66–77. Springer Verlag, 2001.

[8] F. Copty, A. Irron, O. Weissberg, N. Kropp, and G. Kamhi. Efficient debugging in a formal verification environment. *Software Tools for Technology Transfer*, 4:335–348, 2003.

[9] R. Drechsler. *Formal Verification of Circuits.* Kluwer Academic Publishers, 2000.

[10] R. Drechsler. Synthesizing checkers for on-line verification of system-on-chip designs. In *Int'l Symp. Circ. and Systems*, pages IV:748–IV:751, 2003.

[11] R. Drechsler and N. Drechsler. *Evolutionary Algorithms for Embedded System Design.* Kluwer Academic Publisher, 2002.

[12] R. Drechsler and S. Höreth. Gatecomp: Equivalence checking of digital circuits in an industrial environment. In *Int'l Workshop on Boolean Problems*, pages 195–200, 2002.

[13] R. Drechsler and D. Sieling. Binary decision diagrams in theory and practice. *Software Tools for Technology Transfer*, 3:112–136, 2001.

[14] H. Foster, A. Krolnik, and David J. Lacey. *Assertion-Based Design.* Kluwer Academic Publishers, 2003.

[15] S. Hassoun and T. Sasao. *Logic Synthesis and Verification.* Kluwer Academic Publishers, 2001.

[16] P.-H. Ho, T. Shiple, K. Harer, J. Kukula, R. Damiano, V. Bertacco, J. Taylor, and J. Long. Smart simulation using collaborative formal and simulation engines. In *Int'l Conf. on CAD*, pages 120–126, 2000.

[17] Y. Hoskote, T. Kam, P. Ho, and X. Zhao. Coverage estimation for symbolic model checking. In *Design Automation Conf.*, pages 300–305, 1999.

[18] Y.-C. Hsu, B. Tabbara, Y.-A. Chen, and F. Tsai. Advanced techniques for RTL debugging. In *Design Automation Conf.*, pages 362–367, 2003.

[19] P. Johannsen and R. Drechsler. Formal verification on register transfer level – utilizing high-level information for hardware verification. In *IFIP Int'l Conf. on VLSI*, pages 127–132, 2001.

[20] R. Jones. *Symbolic Simulation Methods for Industrial Formal Verification.* Kluwer Academic Publishers, 2002.

[21] A. Kölbl, J. Kukula, and R. Damiano. Symbolic RTL simulation. In *Design Automation Conf.*, pages 47–52, 2001.

[22] Th. Kropf. *Introduction to Formal Hardware Verification.* Springer, 1999.

[23] A. Kuehlmann, M. Ganai, and V. Paruthi. Circuit-based Boolean reasoning. In *Design Automation Conf.*, pages 232–237, 2001.

[24] J. Mohnke, P. Molitor, and S. Malik. Limits of using signatures for permutation independent Boolean comparison. *Formal Methods in System Design: An International Journal*, 2(21):167–191, 2002.

[25] D. Moundanos, J. Abraham, and Y. Hoskote. Abstraction techniques for validation coverage analysis and test generation. *IEEE Trans. on Comp.*, pages 2–14, January 1998.

[26] P. Rashinkar, P. Paterson, and L. Singh. *System-on-a-Chip Verification.* Kluwer Academic Publishers, 2000.

[27] A. Veneris, A. Smith, and M. S. Abadir. Logic verification based on diagnosis techniques. In *ASP Design Automation Conf.*, 2003.

Chapter 1

WHAT SAT-SOLVERS CAN AND CANNOT DO

Eugene Goldberg
Cadence Berkeley Labs, USA
egold@cadence.com

Abstract This chapter consists of two parts. In the first part we show that resolution based SAT-solvers cannot be scalable on real-life formulas unless some extra information about formula structure is known. In the second part we introduce a new way of satisfiability testing that may be used for designing more efficient and "intelligent" SAT-algorithms that will be able to take into account formula structure.

Keywords: Satisfiability problem, resolution, resolution proof complexity, equivalence checking, stable set of points, symmetric CNF formulas

1. Introduction

In the last few years SAT-solvers have considerably improved their performance. As a result, the size of the CNF formulas that can be solved by state-of-the-art SAT-solvers [21, 23, 16, 8] in a reasonable time has dramatically increased. This success has lead to euphoria that reminds many people working in formal verification of early optimism caused by the appearance of BDDs [4]. However, enthusiasts forget that even though SAT-solvers can sometimes solve surprisingly large formulas, they are very far from being scalable (which is the same problem that made people less optimistic about BDDs).

In this chapter, we will try to give a more realistic estimation of the capabilities of SAT-solvers. The chapter is based on the results described in [10, 11, 12] and consists of two parts. The main point of the first part is that a SAT-solver cannot be scalable unless it is provided with some information about the structure of the CNF formula to be tested for satisfiability. In this part, we consider a class of formulas describ-

R. Drechsler (ed.), Advanced Formal Verification, 1-43.

ing equivalence checking of combinational circuits that have a common specification (CS). A CS S of Boolean circuits N_1 and N_2 is just a circuit of multi-valued gates called blocks. Either Boolean circuit is obtained from S by replacing each block of S with its binary implementation. We show that there is a short resolution proof that N_1 and N_2 are equivalent however finding this proof by a deterministic algorithm is most likely infeasible unless a CS of N_1 and N_2 is known. On the one hand, it is bad news. This result means that SAT-algorithms cannot be scalable on equivalence checking CNF formulas (that are important from a practical point of view) even though they have short resolution proofs of unsatisfiability and so are very "easy". On the other hand, this is good news because one can have an efficient algorithm of equivalence checking if a CS of N_1 and N_2 is known. In other words, addressing the question implied by the title of this chapter one can say that SAT-solvers cannot be scalable if no information about high-level structure of formulas is provided.

The result above implies that it is crucial for a SAT-solver to be able to take into account structural properties of formulas. The problem is that the existing SAT-solvers are based on the variable splitting paradigm introduced in the DPLL procedure [7]. During variable splitting a CNF formula is "mutilated" and its subtle structure is usually destroyed. In the second part of this chapter, we introduce a new procedure of satisfiability testing based on the notion of a stable set of points (SSP). It turns out that to prove that a CNF formula F is unsatisfiable it is sufficient to show that F evaluates to 0 (i.e. false) on a set of points called a stable set. In a sense, proving the unsatisfiability of a CNF formula by constructing its SSP can be viewed as "verification" by "simulation".

In general, SSPs are much smaller than the set of all possible assignments but the size of SSPs grows exponentially in the number of variables. So building a monolithic SSP point-by-point can not be used as the basis for designing efficient universal SAT-solvers. We describe two ways of using SSPs. First way is to compute an SSP modulo symmetries of the formula to be tested for satisfiability. In that case, even point-by-point computation of SSPs modulo symmetry can be efficient for highly symmetric formulas. Another way of using SSPs is to replace computing a monolithic SSP with constructing a sequence of much smaller SSPs of "limited" stability. Each such an SSP is stable if "movements" in some directions are forbidden.

2. Hard Equivalence Checking CNF formulas

2.1 Introduction

Since the general resolution system is the basis of almost all practical SAT-solvers, it has been the focus of attention for a long time. In the ground-breaking paper by Haken [13] it was shown that there is a class of CNF formulas for which only exponential size proofs are possible. (In the first part of this chapter we consider only unsatisfiable CNF formulas.) However, the impressive results of state-of-the-art SAT-solvers like Grasp, Sato, Chaff, BerkMin suggest that for the majority of CNF formulas one encounters in practice there should be short resolution proofs of their unsatisfiability. So a natural question to ask is whether the fact that a class of CNF formulas has short resolution proofs means that there is an algorithm that can find these short proofs or proofs that are "close" to them in length. (In complexity theory this question is posed as "whether the general resolution system is automatizable". Studying the automatizability of proof systems was started in [2]. In [18] some results on automatizability of general resolution were obtained.)

The objective of the first part of this chapter is to show that there is a class of CNF formulas that have very short resolution proofs in general resolution that are most likely very hard for a deterministic SAT-algorithm. These formulas specify equivalence checking of Boolean circuits and so they are very important from a practical point of view. This result means that the power of resolution based SAT-solvers is quite limited even for practical formulas that have provably short resolution proofs. The good news is that one can have an efficient SAT-algorithm for solving this class of formulas if some information about the structure of short proofs is provided.

The class of formulas mentioned above describe equivalence checking of circuits having a common "specification". Let N_1 and N_2 be two functionally Boolean circuits with a common specification (CS) S. The CS S is just a circuit of multi-valued gates further referred to as blocks such that N_1 (or N_2) can be obtained from S by replacing each block G of S with its implementation $I_1(G)$ (or $I_2(G)$). The circuit $I_1(G)$ (or $I_2(G)$) implements the multi-output Boolean function obtained from the truth table of G after encoding the values of multi-valued variables with binary codes.

The problem of equivalence checking of N_1 and N_2 can be easily reduced to that of testing the unsatisfiability of a CNF formula (see Section 2.3). Let S consist of n blocks. Let F be a CNF specifying equivalence checking of N_1 and N_2. We show that the unsatisfiability of F can be proven in general resolution in $d * n * 3^{6p}$ resolution steps.

Here d is a constant and p is the size of the largest block G of the CS S (in terms of the number of gates one needs to implement G in N_1 and N_2). In particular, if p is bounded by a constant then we get a class of CNF formulas (in the paper it is denoted by $M(p)$) that has linear size resolution proofs. The parameter p is called the granularity of the specification S.

In spite of the fact that formulas from $M(p)$ have short resolution proofs of unsatisfiability there are good reasons to believe that there does not exist an efficient SAT-algorithm for finding such proofs. Let F be a formula $M(p)$ specifying equivalence checking of circuits N_1 and N_2 with a CS S. Let assume that the CS S is not known. On the one hand, the problem of finding S (or a good approximation of S) is most likely NP-hard. On the other hand, the short resolution proofs mentioned above are closely related to CSs of N_1 and N_2. So given such a short proof of equivalence of N_1 and N_2 one could recover a "good" CS from this proof. Hence the existence of an efficient procedure for finding a short proof of equivalence would mean that there is an efficient algorithm for solving an NP-hard problem.

As we mentioned above the good news is that a formula F of $M(p)$ can be efficiently solved by a deterministic algorithm if some extra information is provided. This extra information is a CS S of N_1 and N_2 whose equivalence checking the formula F specifies. (Namely, one just needs to know the assignment of gates of N_1 and N_2 to blocks of S. No other information about S is needed. In particular, one needs neither any knowledge of the functionality of blocks of S nor the knowledge of binary encodings used when producing N_1 and N_2 from S.) We formulate a specification aware algorithm of checking the unsatisfiability formulas from $M(p)$ that has the same complexity as resolution proofs. That is it solves the formulas of $M(p)$ in linear time.

The first part of this chapter is structured as follows. In Section 2.2 we introduce the notion of a CS of Boolean circuits that plays a key role in the following exposition. Section 2.3 describes a common way of reducing equivalence checking to SAT. In Section 2.4 we introduce a class $M(p)$ of CNF formulas encoding equivalence checking of Boolean circuits with a CS of granularity p. We also describe the general resolution proof system. Section 2.5 describes computation of existentially implied functions that is used in Section 2.6. In the latter, we proof the main result of the first part of this chapter about the complexity of formulas from $M(p)$ in general resolution. In Section 2.7 and 2.8 we discuss the complexity of formulas $M(p)$ for deterministic resolution based algorithms. In Section 2.7 we give reasons why formulas from $M(p)$ should be hard for deterministic SAT-algorithms that do not have any knowledge of a CS of the circuits checked for equivalence. In Section 2.8 we describe an efficient resolution based SAT-algorithm for equivalence checking of circuits with a known CS. In Section 2.9 we show experimentally that formulas from $M(p)$ are hard for existing SAT-solvers while a specification aware algorithm easily solves them. In Section 2.10 some conclusions are made.

2.2 Common Specification of Boolean Circuits

In this section, we introduce the notion of a common specification of Boolean circuits. Let S be a combinational circuit of multi-valued blocks (further referred to as a ***specification***) specified by a directed acyclic graph H. (An example of specification is shown in Fig. 1.1*a*.) The sources and sinks of H correspond to primary inputs and outputs of S. Each non-source node of H corresponds to a multi-valued block computing a multi-valued function of multi-valued arguments. Each node n of H is associated with a ***multi-valued variable*** V. If n is a source of H , then the corresponding variable specifies values taken by the corresponding primary input of S. If n is a non-source node of S then the corresponding variable describes the values taken by the output of the block specified by n. If n is a source (respectively a sink), then the corresponding variable is called a ***primary input variable*** *(*respectively ***primary output variable)***. We will use the notation C=G(A,B) to indicate that a) the output of a block G is associated with a variable C; b) the function computed by the block G is G(A,B); c) only two nodes of H are connected to the node n in H and these nodes are associated with variables A and B.

Denote by $D(A)$ the ***domain*** of a variable A associated with a node of H. The value of $|D(A)|$ is called the ***multiplicity*** of A. If the multiplicity of every variable A of S is equal to 2 then S is a ***Boolean circuit***.

Now we describe how a Boolean circuit N can be produced from a specification S by encoding the multi-valued variables. Let $D(A) = \{a_1, \dots, a_t\}$ be the domain of a variable A of S. Denote by $q(A)$ ***a Boolean encoding*** of the values of $D(A)$ that is a mapping $q : D(A) \rightarrow \{0,1\}^m$. Denote by *length*$(q(A))$ the number of bits in q that is the value of m. The value of $q(a_i)$, $a_i \in D(A)$ is called the ***code*** of a_i. Given an encoding q of length m of a variable A associated with a block of S, denote by $v(A)$ the set of m ***coding Boolean variables***.

Example 1.1 *Let B be a multi-valued variable and $D(B) = \{b_1, b_2, b_3, b_4\}$. Then the multiplicity of the variable B is 4. Let a mapping q be specified by the following expressions $q(b_1) = 01, q(b_2) =$*

$11, q(b_3) = 10, q(b_4) = 00$. Then q specifies an encoding of the values of B of length($q(B)$) equal to 2. The set of coding variables $v(B) = \{q_1, q_2\}$ consists of two Boolean variables. The Boolean vector 01 *where $q_0 = 0, q_1 = 1$ is the code of b_1 under the encoding q.*

In the following exposition we make the assumptions below.

Assumption 1.1 *Each gate of a Boolean circuit and each block of a specification has two inputs and one output.*

Assumption 1.2 *The multiplicity of each primary input (or output) variable of a specification is a power of 2.*

Assumption 1.3 *If V is a primary input (or output) variable of a specification, then length($q(A)$) $= log_2(|D(A)|)$*

Assumption 1.4 *If a_1 and a_2 are values of a variable A of a specification and $a_1 \neq a_2$, then $q(a_1) \neq q(a_2)$.*

Assumption 1.5 *If A and B are two different variables of a specification, then $v(A) \cap v(B) = \varnothing$.*

Remark 1.1 *From Assumptions 1.2, 1.3 and 1.4 it follows that if A is a primary input (or output) variable, a mapping $q : D(A) \rightarrow \{0,1\}^m$ is bijective. In particular, any assignment to the variables of $v(A)$ is a code of some value $a \in D(A)$.*

Definition 1.1 *Given a Boolean circuit I, denote by Inp(I) (respectively Out(I)) the set of variables associated with primary inputs (respectively primary outputs) of I.*

Definition 1.2 *Let X_1 and X_2 be sets of Boolean variables and $X_2 \subseteq X_1$. Let y be an assignment to the variables of X_1. Denote by proj(y, X_2) the* ***projection*** *of y on X_2 i.e. the part of y that consists of the assignments to the variables of X_2.*

Example 1.2 *Let $X_1 = \{x_1, x_2, x_3, x_4, x_5\}$ and $X_2 = \{x_1, x_3, x_5\}$ that is $X_2 \subseteq X_1$. Let y be the assignment $(x_1 = 0, x_2 = 1, x_3 = 1, x_4 = 0, x_5 = 0)$ to the variables of X_1. Then proj(y, X_2) is equal to $(x_1 = 0, x_3 = 1, x_5 = 0)$.*

Definition 1.3 *Let C=$G(A,B)$ be a block of specification S. Let $q(A)$, $q(B)$, $q(C)$ be encodings of variables A,B, and C respectively. A Boolean* *circuit I is said to* ***implement the block G*** *if the following three conditions hold:*

- *The set Inp(I) is a subset of* $v(A) \cup v(B)$.
- *The set Out(I) is equal to* $v(C)$.
- *If the set of values assigned to* $v(A)$ *and* $v(B)$ *form codes* $q(a)$ *and* $q(b)$ *respectively where* $a \in D(A)$, $b \in D(B)$, *then* $I(z')=q(c)$ *where* z' *is the projection of the assignment* $z=(q(a),q(b))$ *on Inp(I),* $I(z')$ *is the value taken by I at* z', *and* $c=G(a,b)$

Example 1.3 *In Fig. 1.1a a specification of three blocks is shown. The functionality of two different implementations of the block* $C=G_1(A,B)$ *(Fig 1.1b) is shown in Fig. 1.1c and 1.1d. Here* $D(A)=\{a_0,a_1\}$, $D(B)=\{b_0,b_1,b_2,b_3\}$ *and* $D(C)=\{c_0,c_1,c_2\}$. *Since A and B are primary input variables they are encoded with a minimum length code and* $q_1(A)=q_2(A)$ *and* $q_1(B)=q_2(B)$ *where* $q_1(a_0)=0$, $q_1(a_1)=1$, $q_1(b_0)=00$, $q_1(b_1)=01$, $q_1(b_2)=10$, $q_1(b_3)=11$. *Finally, the encodings* $q_1(C)$ *and* $q_2(C)$ *are* $q_1(c_0)=00$, $q_1(c_1)=10$, $q_1(c_2)=01$ *and* $q_2(c_0)=100$, $q_2(c_1)=010$, $q_2(c_2)=001$.

Remark 1.2 *The reason why Inp(I) in Definition 1.3 may not include all the variables of* $v(A)$ *and/or* $v(B)$ *is that the function* $G(A,B)$ *may not distinguish some values of A or B. (*$G(A,B)$ *does not distinguish, say, values* $a_1, a_2 \in D(A)$, *if for any* $b \in D(B)$, $G(a_1,b)=G(a_2,b)$.*) So to implement* $G(A,B)$ *the circuit I may need only a subset of variables of* $v(A) \cup v(B)$. *This situation is illustrated in Fig. 1.2. Due to the fact that some values of the variable C are indistinguishable by* G_2, *only two outputs of the implementation block* $I(G_1)$ *(out of the three) are connected to the inputs of* $I(G_2)$. *This said, henceforth, for the sake of simplicity, we will write* $I(q(a),q(b))$ *meaning* $I(q'(a),q'(b))$, $q'(a)=proj(q(a),Inp(I))$ *and* $q'(b)=proj(q(b),Inp(I))$.

Definition 1.4 *Let S be a multi-valued circuit. A Boolean circuit N is said to* ***implement the specification S****, if it is built according to the following two rules.*

- *Each block G of S is replaced with an implementation I of G.*
- *Let the output of block* G_1 *(specified by variable C) be connected to an input of block* G_2 *(specified by the same variable C) in S. Then the outputs of the circuit* I_1 *implementing* G_1 *are properly*

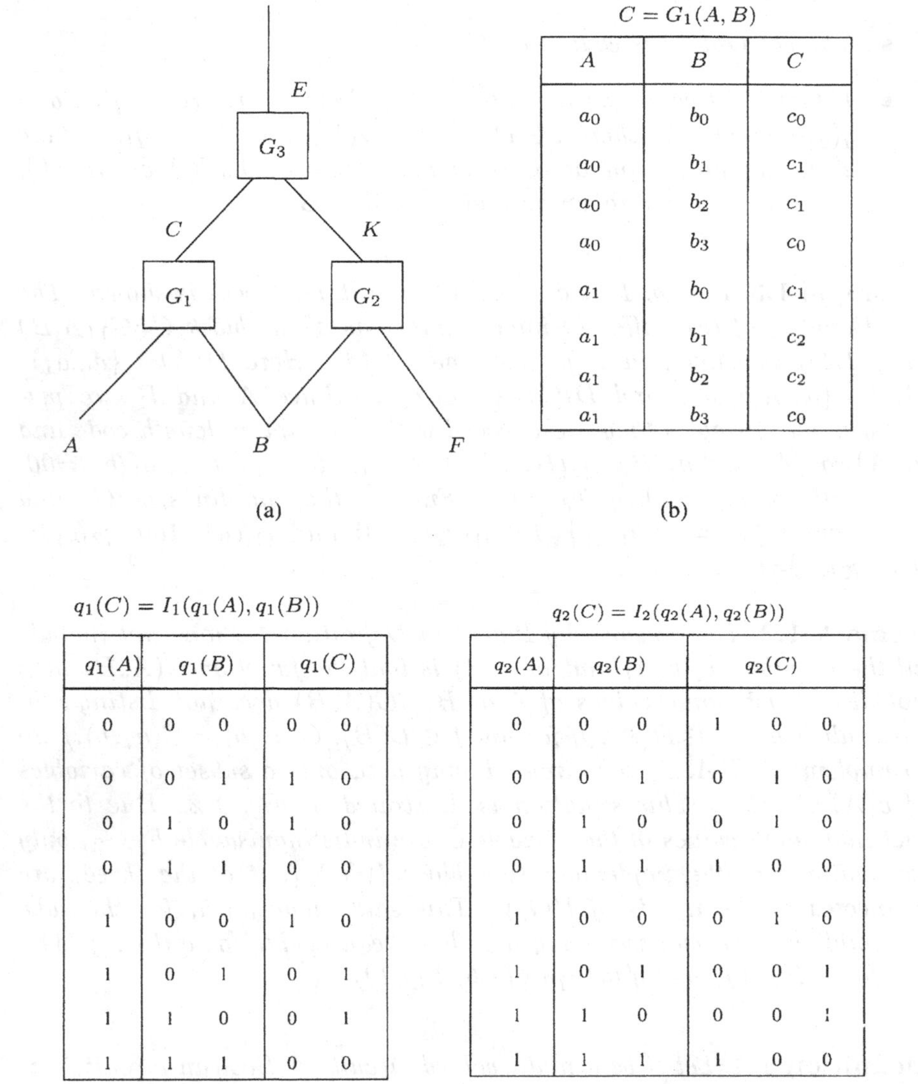

$C = G_1(A, B)$

A	B	C
a_0	b_0	c_0
a_0	b_1	c_1
a_0	b_2	c_1
a_0	b_3	c_0
a_1	b_0	c_1
a_1	b_1	c_2
a_1	b_2	c_2
a_1	b_3	c_0

(b)

$q_1(C) = I_1(q_1(A), q_1(B))$

$q_1(A)$	$q_1(B)$		$q_1(C)$	
0	0	0	0	0
0	0	1	1	0
0	1	0	1	0
0	1	1	0	0
1	0	0	1	0
1	0	1	0	1
1	1	0	0	1
1	1	1	0	0

(c)

$q_2(C) = I_2(q_2(A), q_2(B))$

$q_2(A)$	$q_2(B)$		$q_2(C)$		
0	0	0	1	0	0
0	0	1	0	1	0
0	1	0	0	1	0
0	1	1	1	0	0
1	0	0	0	1	0
1	0	1	0	0	1
1	1	0	0	0	1
1	1	1	1	0	0

(d)

Figure 1.1. A specification and the functionality of two implementations of block G_1

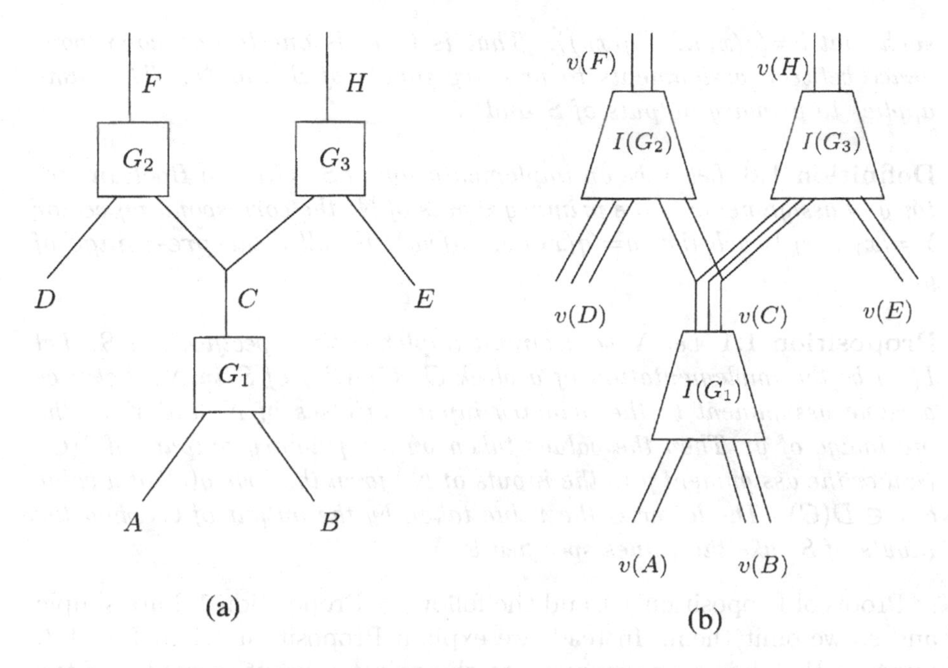

Figure 1.2. A specification and its implementation

connected to inputs of circuit I_2 implementing G_2. Namely, the primary output of I_1 specified by a Boolean variable $q_i \in v(C)$ is connected to the input of I_2 specified by the same variable of $v(C)$ if $q_i \in Inp(I_2)$.

Fig. 1.2 gives an example of a specification (Fig. 1.2*a*) and its implementation (Fig. 1.2*b*).

Remark 1.3 *Let N be an implementation of a specification S. Let p be the largest number of gates used in an implementation of a multi-valued block of S in N. We will say that S is a specification of* **granularity** *p for N.*

Definition 1.5 *The* **topological level** *of a block G in a specification S is the length of the longest path from a primary input of S to G. (The length of a path is measured in the number of blocks on it. The topological level of a primary input is assumed to be 0.) Denote by* **level(G)** *the topological level of G in S.*

Remark 1.4 *Let N be an implementation of a specification S. From Remark 1.1 it follows that for any value assignment h to the input variables of N there is a unique set of values $(x_1,\ldots,x_k)$, where $x_i \in D(X_i)$*

such that h=$(q(x_1),\ldots,q(x_k))$. That is there is one-to-one correspondence between assignments to primary inputs of S and N. The same applies to primary outputs of S and N.

Definition 1.6 *Let N be an implementation of S. Given a Boolean vector y of assignments to the primary inputs of N, the corresponding vector Y=$(x_1,..,x_k)$ such that y=$(q(x_1),\ldots,q(x_k))$ is called the* ***pre-image*** *of y.*

Proposition 1.1 *Let N be a circuit implementing specification S. Let $I(G)$ be the implementation of a block C=$G(A,B)$ of S in N. Let y be a value assignment to the primary input variables of N and Y be the pre-image of y. Then the values taken by the primary outputs of $I(G)$ (under the assignment y to the inputs of N) form the code $q(c)$ of a value c, $c \in D(C)$. The latter is the value taken by the output of G when the inputs of S take the values specified by Y.*

Proofs of Proposition 1.1 and the following Proposition 1.2 are simple and so we omit them. Instead, we explain Proposition 1.1 in Fig. 1.3. Suppose that y is an assignment to the primary input variables of the Boolean circuit (Fig. 1.3a) that is an implementation of the specification shown in Fig. 1.3b. According to Remark 1.4, y can be represented as $(q(a), q(b), q(d), q(e))$ where a, b, d, e are values of the variables A, B, D, E of the specification respectively. The pre-image of y is the vector $Y = (a, b, d, e)$. Then the outputs of gates G_1,G_2 and G_3 take values $c = G_1(a, b)$,$f = G_2(d, c)$ and $h = G_3(c, e)$ respectively. Since $I(G_1)$,$I(G_2)$ and $I(G_3)$ are implementations of G_1,G_2,G_3 respectively, their outputs take values $q(c)$,$q(f)$ and $q(h)$ respectively.

Proposition 1.2 *Let N_1, N_2 be circuits implementing a specification S. Let each primary input (or output) variable X of S have the same encoding in N_1 and N_2. Then Boolean circuits N_1 and N_2 are functionally equivalent.*

Definition 1.7 *Let N_1, N_2 be two functionally equivalent Boolean circuits. Let N_1, N_2 implement a specification S so that for every primary input (output) variable X encodings $q_1(X)$ and $q_2(X)$ (used when producing N_1 and N_2 respectively) are identical. Then S is called* ***a common specification*** *(CS) of N_1 and N_2.*

Assumption 1.6 *Let S be a CS of N_1,N_2 and C be a variable of S. We will assume that $v_1(C)$ =$v_2(C)$ if C is a primary input variable and $v_1(C) \cap v_2(C) = \varnothing$ otherwise.*

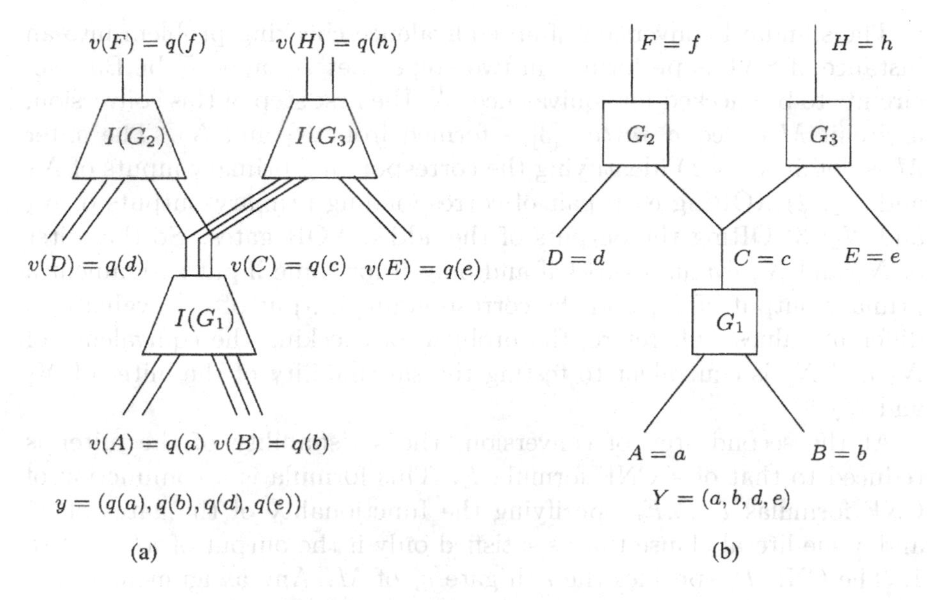

Figure 1.3. An illustration to Proposition 1.1

Definition 1.8 *Let S be a CS of N_1,N_2. Let p_1 (respectively p_2) be the granularity of S with respect to N_1 (respectively N_2). Then we will say that S is a CS of N_1,N_2 of* ***granularity*** $p = max(p_1,p_2)$.

Definition 1.9 *Given two functionally equivalent Boolean circuits N_1, N_2, S is called the* ***finest common specification*** *if it has the smallest granularity p among all the CSs of N_1 and N_2.*

2.3 Equivalence Checking as SAT

In this section, we recall a common way of reducing equivalence checking to the satisfiability problem.

Definition 1.10 *A disjunction of literals of Boolean variables not containing two literals of the same variable is called a* ***clause****. A conjunction of clauses is called a* ***conjunctive normal form*** *(CNF).*

Definition 1.11 *Given a CNF F,* ***the satisfiability problem*** *(SAT) is to find a value assignment to the variables of F for which F evaluates to 1 (also called a* ***satisfying assignment****) or to prove that such an assignment does not exist. A clause K of F is said to be* ***satisfied*** *by a value assignment y if $K(y) = 1$. If $K(y) = 0$, the clause K is said to be* ***falsified*** *by y.*

The standard conversion of an equivalence checking problem into an instance of SAT is performed in two steps. Let N_1 and N_2 be Boolean circuits to be checked for equivalence. At the first step of this conversion, a circuit M called ***a miter*** [3] is formed from N_1 and N_2. The miter M is obtained by 1) identifying the corresponding primary inputs of N_1 and N_2; 2) XORing each pair of corresponding primary outputs of N_1 and N_2; 3) ORing the outputs of the added XOR gates. So the miter of N_1 and N_2 evaluates to 1 if and only if for some input assignment a primary output of N_1 and the corresponding output of N_2 evaluate to different values. Therefore, the problem of checking the equivalence of N_1 and N_2 is equivalent to testing the satisfiability of the miter of N_1 and N_2.

At the second step of conversion, the satisfiability of the miter is reduced to that of a CNF formula F. This formula is a conjunction of CNF formulas F_1,..,F_n specifying the functionality of the gates of M and a one-literal clause that is satisfied only if the output of M is set to 1. The CNF F_i specifies the i-th gate g_i of M. Any assignment to the variables of F_i that is inconsistent with the functionality of g_i falsifies a clause of F_i (and vice versa, a consistent assignment satisfies all the clauses of F_i.) For instance, the AND gate $y{=}x_1x_2$ is specified by the following three clauses $\sim x_1 \vee \sim x_2 \vee y$, $x_1 \vee \sim y$, $x_2 \vee \sim y$.

2.4 Class $M(p)$ and general resolution

In this short section we formally define the class of equivalence checking formulas we consider in the first part of this chapter. Besides, we describe the general resolution system.

Definition 1.12 *Given a constant p, a CNF formula F is a member of the* ***class M(p)*** *if and only if it satisfies the following two conditions.*

- *F is the CNF formula (obtained by the procedure described in Section 2.3) specifying the miter of a pair of functionally equivalent circuits* N_1,N_2.
- N_1,N_2 *have a CS of granularity* p' *where* $p' \leq p$.

Definition 1.13 *Let K and K′ be clauses having opposite literals of a variable (say variable x) and there is only one such variable. The* ***resolvent*** *of K , K′ in variable x is the clause that contains all the literals of K and K′ but the positive (i.e. literal x) and negative (i.e. literal* $\sim x$*) literals of x. The operation of producing the resolvent of K and K′ is called* ***resolution****.*

Definition 1.14 ***General resolution*** *is a proof system of propositional logic that has only one inference rule. This rule is to resolve two existing clauses to produce a new one. Given a CNF formula F, a proof $L(F)$ of unsatisfiability of F in the general resolution system consists of a sequence of resolutions resulting in the derivation of an* ***empty clause*** *(i.e. a clause without literals).*

General resolution is complete. This means that given an unsatisfiable formula F there is always a sequence of resolutions $L(F)$ in which an empty clause is derived.

2.5 Computation of existentially implied functions

In this section, we introduce the notion of existentially implied functions that is used in Section 2.6 in the definitions of filtering and correlation functions.

Definition 1.15 *Let $F(X_1, X_2)$ be a Boolean function where X_1 and X_2 are sets of Boolean variables. The function $H(X_2)$ is called* ***existentially implied*** *by F if*

- $F(X_1,X_2) \rightarrow H(X_2)$
- *if $H(z)=1$ where z is an assignment to the variables of X_2, then there is an assignment y to the variables of X_1 such that $F(y,z)=1$.*

Remark 1.5 *Given a function $F(X_1,X_2)$, the function $H(X_2)$ existentially implied by F is unique. It can be obtained from F by existentially quantifying away the variables of X_1.*

Proposition 1.3 *Let $F(X_1, X_2)$ and $H(X_2)$ be CNF formulas where $H(X_2)$ consists of all the clauses depending only on variables from X_2 that can be derived from $F(X_1,X_2)$ by resolution. Then $H(X_2)$ is existentially implied by $F(X_1,X_2)$.*

Proof. The CNF $F(X_1,X_2)$ implies $H(X_2)$ because each clause of H is implied by F since it is derived by resolution. Assume that H is not existentially implied by F. Then there is an assignment z to the variables of X_2 such that $H(z)=1$ and for any assignment y to the variables of X_1, $F(y,z)=0$. However, this means that F implies a clause K depending only on variables of X_2 such that $K(z)=0$. Since K should be in H, then $H(z)$ should be equal to 0, which leads to a contradiction.

Definition 1.16 *Let F be a set of clauses. Denote by* ***supp(F)*** *the set of variables whose literals occur in clauses of F.*

To estimate the complexity of obtaining the function existentially implied by F in general resolution, we need the following proposition.

Proposition 1.4 *Let F be a set of clauses that implies a clause K. Then there is a sequence of at most $3^{|supp(F)|}$ resolution steps that results in the derivation of the clause K or a clause that implies K.*

Proof. Denote by F' the formula that is obtained from F by making the assignments that set the literals of K to 0 (and removing the satisfied clauses and the literals set to 0). It is not hard to see that F' is unsatisfiable since it implies an empty clause. So there is a resolution proof $L(F')$ that results in deducing an empty clause. Then by replacing each clause of F' involved in $L(F')$ with its "parent" clause from F we get a sequence of resolutions resulting in deducing either the clause K or a clause that implies K. The number of resolvents in $L(F')$ cannot be more than $3^{|supp(F')|}$ (which is the total number of clauses of $|supp(F')|$ variables) and so it cannot be more than $3^{|supp(F)|}$.

Remark 1.6 *From Propositions 1.3 and 1.4 it follows that given a CNF $F(X_1, X_2)$ one can obtain the function $H(X_2)$ existentially implied by F in no more than $3^{|supp(F)|}$ resolution steps.*

2.6 Equivalence Checking in General Resolution

In this section, we prove some results about the complexity of formulas of the class $M(p)$ (describing equivalence checking of circuits with a CS of granularity p) in general resolution. The main idea of the proof is that if S is a CS of N_1 and N_2, then their equivalence checking reduces to computing filtering and correlation functions. For each variable C of the specification S one needs to compute filtering functions $Ff(v_1(C))$,$Ff(v_2(C))$ and the correlation function $Cf(v_1(C), v_2(C))$. Here $v_1(C)$ (respectively $v_2(C)$) are coding variables of the encoding $q_1(C)$ (respectively $q_2(C)$) used when obtaining the implementation N_1 (respectively N_2).

The three main properties of these functions are that

- They can be built based only on the information about the topology of S and about "assignment" of gates of N_1 and N_2 to blocks of S.
- Filtering functions and correlation functions corresponding to primary input variables of the specification are constants.
- Filtering and correlation functions for a variable C specifying the output of a block $G(A, B)$ can be computed "locally" from filtering

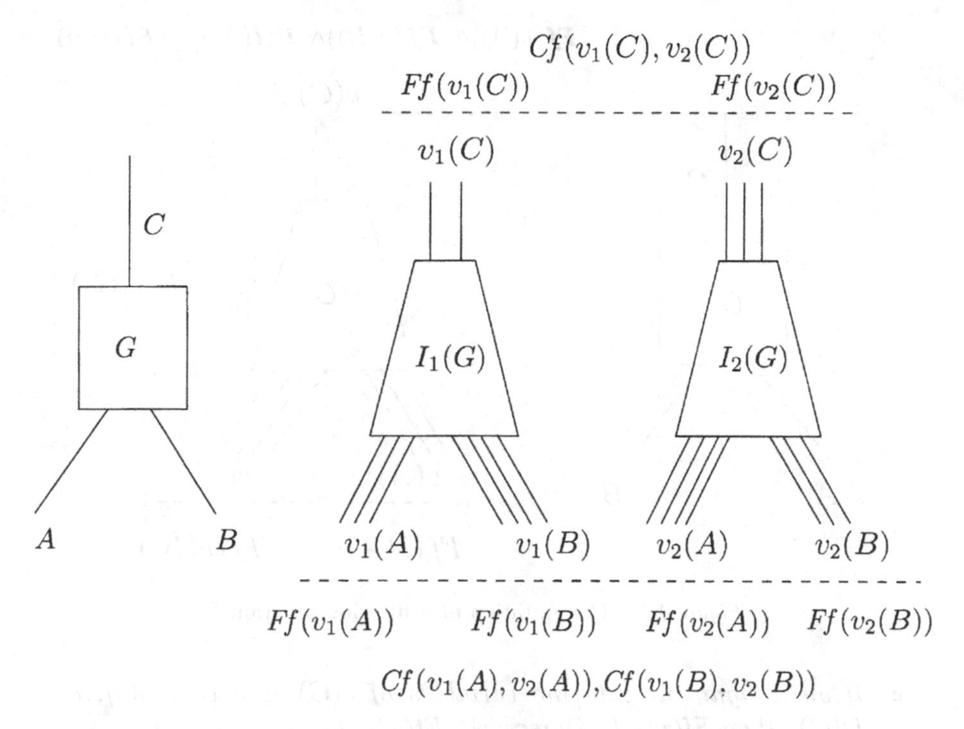

Figure 1.4. Computation of filtering and correlation functions

and correlation functions of variables A and B and CNFs specifying implementations $I_1(G)$ and $I_2(G)$. So these functions can be computed in topological order starting with inputs and proceeding to outputs.

A general scheme for the computation of filtering and correlation functions is shown in Fig. 1.4. To compute the filtering functions $Ff(v_1(C))$ and $Ff(v_2(C))$ and the correlation function $Cf(v_1(C), v_2(C))$ one needs to know filtering functions $Ff(v_1(A))$,$Ff(v_2(A))$,$Ff(v_1(B))$, $Ff(v_2(B))$ and correlation functions $Cf(v_1(A), v_2(A))$,$Cf(v_1(B), v_2(B))$.

In this chapter, we consider computation of filtering and correlation functions (represented as CNF formulas) in the general resolution proof system. However, one can use other ways of computing these functions, for example, employing BDDs[4].

Definition 1.17 *Let N be an implementation of a specification S. Let C be a variable of S. A function Ff is called* ***a filtering function*** *if:*

- *$supp(Ff) \subseteq v(C)$.*

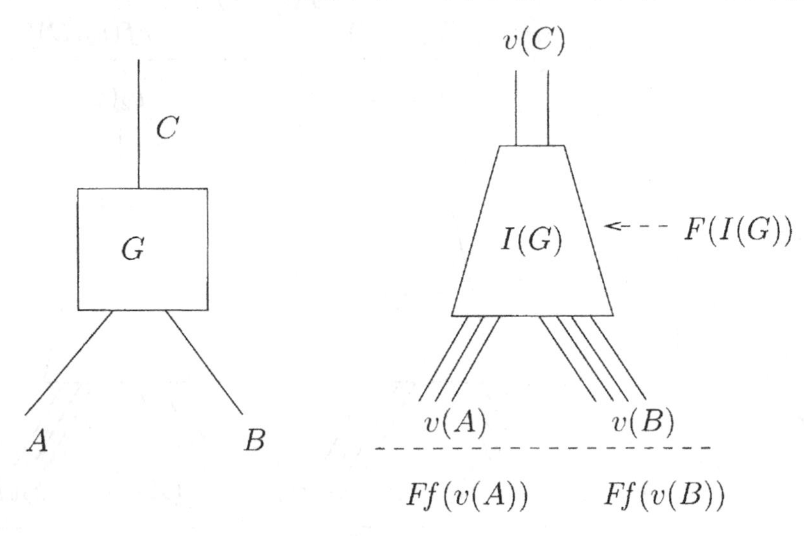

Figure 1.5. Computation of a filtering function

- *If an assignment z to the variables of $v(C)$ is a code $q(c)$, $c \in D(C)$, then $Ff(z)$=1. Otherwise, $Ff(z)$=0.*

Remark 1.7 *If C is a primary input variable of S, then $Ff(v(C)) \equiv 1$. Indeed, as it follows from Remark 1.1, any assignment to $v(C)$ is the code of a value $c \in D(C)$.*

Proposition 1.5 *Let N be an implementation of a specification S. Let C=$G(A,B)$ be a block of S. Let F be the CNF formula specifying N built as described in Section 2.3 and $F(I(G))$ be the part of F specifying the implementation $I(G)$ of G in N. Then P existentially implies $Ff(v(C))$ where P=$Ff(v(A)) \wedge Ff(v(B)) \wedge F(I(G))$.*

Proof. To make it easier for the reader to "visualize" the proof, we illustrate the proposition in Fig. 1.5. To prove that $P \rightarrow Ff(v(C))$ one needs to show that any assignment that sets P to 1 also sets $Ff(v(C))$ to 1. It is not hard to see that the support of all the functions of the expression $P \rightarrow Ff(v(C))$ is a subset of $supp(F(I(G)))$. Let h=(x,y,z) be an assignment that sets P to 1 where x,y,z are assignments to the variables from $v(A)$,$v(B)$ and $v(C)$ respectively. Then h has to set to 1 the functions $Ff(v(A))$, $Ff(v(B)$, $F(I(G))$. Since h sets $Ff(v(A))$ to 1, then x=$q(a)$, $a \in D(A)$. Since h sets $Ff(v(B))$ to 1, then y=$q(b)$, $b \in$

$D(B)$. So $h = (q(a), q(b), z)$. To set to 1 $F(I(G))$, assignment z has to be equal to $q(c)$, where c=$G(a,b)$. Then h sets $Ff(v(C))$ to 1.

Assume that $Ff(v(C))$ is not existentially implied by P. Then there exists an assignment z=$q(c)$, $c \in D(C)$ such that $Ff(z)$=1 and for any assignments x and y to the variables of $v(A)$ and $v(B)$ respectively, $P(x,y,z)$=0. However, $P(q(a), q(b), z) = 1$ where a and b are values of A and B such that $G(a,b)$=c, which leads to a contradiction.

Definition 1.18 *Let S be a CS of circuits N_1 and N_2 and C be a variable of S. A function Cf is called* ***a correlation function*** *for encodings q_1 and q_2 of the values of C (used when producing N_1 and N_2) if :*

- *supp(Cf) $\subseteq v_1(C) \cup v_2(C)$.*
- *Cf(z_1, z_2)=1 for any assignment z_1 to $v_1(C)$ and z_2 to $v_2(C)$ such that z_1=$q_1(c)$ and z_2=$q_2(c)$ where $c \in D(C)$. Otherwise Cf(z_1, z_2)=0.*

Remark 1.8 *If C is a primary input variable of S, then Cf($v_1(C)$, $v_2(C)$) $\equiv$ 1. Indeed, as it follows from Remark 1.1, any assignment to $v_1(C)$ or $v_2(C)$ is the code of a value $c \in D(C)$. Besides, from the definition of CS it follows that $q_1(C)$=$q_2(C)$. Finally, from Assumption 1.6 it follows that $v_1(C) = v_2(C)$. So any assignment (x, y) to the variables of $v_1(C)$,$v_2(C)$ can be represented as ($q_1(c)$,$q_2(c)$), $c \in D(C)$.*

Proposition 1.6 *Let S be a CS of circuits N_1,N_2. Let $C = G(A, B)$ be a block of S. Let F be the CNF formula specifying the miter of N_1,N_2 built as described in Section 2.3. Let $F(I_1(G))$ and $F(I_2(G))$ be the part of F specifying the implementation $I_1(G)$ and $I_2(G)$ of G in N_1 and N_2 respectively. Then P existentially implies Cf($v_1(C), v_2(C)$) where*

- *P = Filtering $\wedge$ Correlation $\wedge$ Implementation*
- *Filtering = Ff($v_1(A)$) $\wedge$ Ff($v_1(B)$) $\wedge$ Ff($v_2(A)$) $\wedge$ Ff($v_2(B)$)*
- *Correlation = Cf($v_1(A)$,$v_2(A)$) $\wedge$ Cf($v_1(B)$,$v_2(B)$)*
- *Implementation = $F(I_1(G)) \wedge F(I_2(G))$.*

Proof. To make it easier for the reader to "visualize" the proof, we illustrate the proposition in Fig. 1.6. To prove that P implies *Cf* ($v_1(C)$, $v_2(C)$) one needs to show that any assignment that sets P to 1 also sets *Cf* ($v_1(C)$,$v_2(C)$) to 1. It is not hard to see that the support of all the functions of the expression $P \rightarrow$ *Cf* ($v_1(C)$,$v_2(C)$) is a subset of

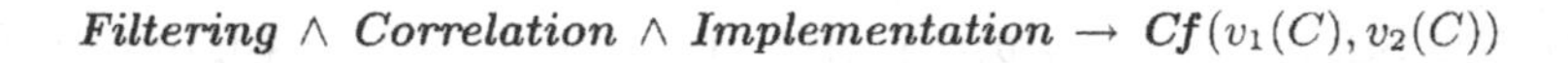

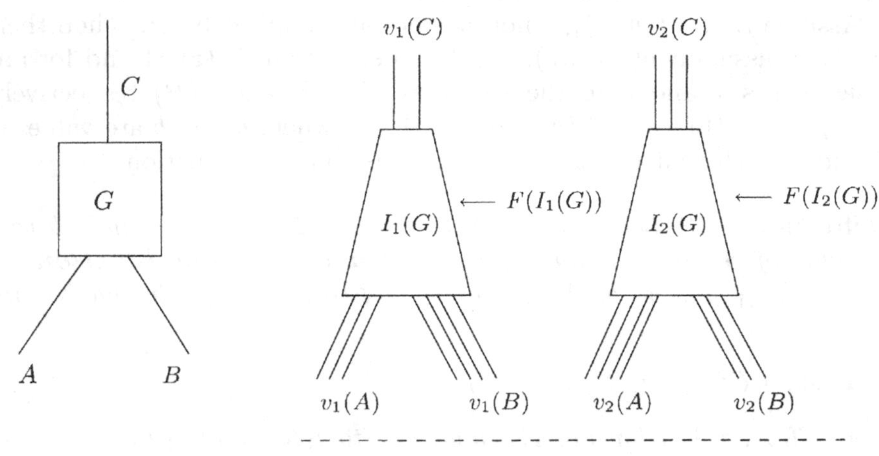

Filtering $= Ff(v_1(A)) \wedge Ff(v_1(B)) \wedge Ff(v_2(A)) \wedge Ff(v_2(B))$

Correlation $= Cf(v_1(A), v_2(A)) \wedge Cf(v_1(B), v_2(B))$

Implementation $= F(I_1(G)) \wedge F(I_2(G))$

Figure 1.6. Computation of a correlation function

$supp(F(I_1(G)) \cup supp(F(I_2(G))$. Let $h{=}(x_1, x_2, y_1, y_2, z_1, z_2)$ be an assignment that sets P to 1 where $x_1, x_2, y_1, y_2, z_1, z_2$ are assignments to $v_1(A)$, $v_2(A)$, $v_1(B)$, $v_2(B)$, $v_1(C)$, $v_2(C)$ respectively. Then h has to set to 1 all the functions the conjunction of which forms P. Since h has to set the function *Filtering* to 1, then $x_1{=}q_1(a_1)$, $x_2{=}q_2(a_2)$ where a_1,$a_2 \in D(A)$ and $y_1{=}q_1(b_1)$, $y_2{=}q_2(b_2)$, where b_1,$b_2 \in D(B)$. So $h{=}(q_1(a_1),q_2(a_2), q_1(b_1),q_2(b_2), z_1, z_2)$. Since h sets the function *Correlation* to 1, then a_1 has to be equal to a_2 and b_1 has to be equal to b_2. So h can be represented as $(q_1(a),q_2(a), q_1(b),q_2(b), z_1, z_2)$ where $a \in D(A)$ and $b \in D(B)$. Since h sets the function *Implementation* to 1, then z_1 has to be equal to $q_1(c)$, $c{=}G(a,b)$ and z_2 has to be equal to $q_2(c)$. So h is equal to $(q_1(a),q_2(a),q_1(b),q_2(b),q_1(c),q_2(c))$ and hence it sets the correlation function $Cf(v_1(C),v_2(C))$ to 1.

Assume that $Cf(v_1(C),v_2(C))$ is not existentially implied by P. Then there exists an assignment $z_1{=}q_1(c)$, $z_2{=}q_2(c)$ to the variables of $v_1(C)$ and $v_2(C)$ respectively such that $Cf(z_1, z_2){=}1$ and for any assignment x_1, x_2, y_1, y_2 to the variables of $v_1(A)$, $v_2(A)$, $v_1(B)$, $v_2(B)$ respectively, $P(x_1, x_2, y_1, y_2, z_1, z_2){=}0$. However, $P(q_1(a), q_2(a), q_1(b), q_2(b), z_1, z_2){=}1$ where a, b are the values of A and B respectively for which $c{=}G(a,b)$. This leads to a contradiction.

Proposition 1.7 *Let F be a formula of $M(p)$ specifying the miter of circuits N_1,N_2 obtained from a CS S of granularity p. The unsatisfiability of F can be proven by a resolution proof of no more than $d*n*3^{6p}$ resolution steps where n is the number of blocks in S and d is a constant.*

Proof. From Proposition 1.5 and Proposition 1.6 it follows that one can deduce correlation and filtering functions for all the variables of S starting with blocks of topological level 1 and proceeding in topological order. Indeed, let C=$G(A,B)$ be a block of topological level 1. Then A and B are primary input variables and the filtering and correlation functions for them are known (they are tautologies). Then $Ff(v_1(C))$ and $Ff(v_2(C))$ are existentially implied by $F(I_1(G))$ and $F(I_2(G))$ respectively. According to Proposition 1.5 $Ff(v_1(C))$ (respectively $Ff(v_2(C))$) can be derived by resolving clauses of $F(I_1(G))$ (respectively $F(I_2(G))$). Similarly, the correlation function $Cf(v_1(C),v_2(C))$ is existentially implied by $F(I_1(G)) \wedge F(I_2(G))$. So it can be derived from the latter by resolution. After filtering and correlation functions are computed for all the variables of level 1, the same procedure can be applied to variables of topological level 2 and so on. If S consists of n blocks, then in n steps one can deduce correlation functions for the primary output variables of S. At each step two filtering and one correlation function are computed for a variable C=$G(A,B)$ of S. The complexity of this step is no more than 3^{6p}. Indeed, the support of all functions mentioned in Proposition 1.5 and Proposition 1.6 needed for computing $Ff(v_1(C))$, $Ff(v_2(C))$ and $Cf(v_1(C),v_2(C))$ is a subset of E=$supp(F(I_1(G))) \cup supp(F(I_2(G)))$. The total number of gates in $I_1(G)$ and $I_2(G)$ is bounded by $2p$, each gate having 2 inputs and 1 output. So the total number of variables in E cannot be more than $6p$. Then according to Remark 1.6 in no more than 3^{6p} steps one can deduce CNFs $Ff(v_1(C))$, $Ff(v_2(C))$ and $Cf(v_1(C),v_2(C))$. Then the total number of resolution steps one needs to deduce correlation functions for primary output variables of S is bounded by $n*3^{6p}$.

Now we show that from the correlation functions for primary output variables of S one can deduce an empty clause in the number of resolution steps linear in $n*p$. Let C be a primary output variable specifying the output of a block G of N. Let $I_1(G)$ and $I_2(G)$ be the implementations of G in N_1 and N_2 respectively. Let $|D(C)| = 2^k$ (By Assumption 1.2 the multiplicity of C is a power of 2.) Then $length(q_1(C))$= $length(q_2(C))$=k. (By Assumption 1.3 values of C are encoded by a minimal length encoding.)

Now we show that there is always a correlation function $Cf(v_1(C),$ $v_2(C))$ specified by the CNF consisting of k pairs of two literal clauses specifying the equivalence of corresponding outputs of $I_1(G)$ and $I_2(G)$. Let f_1 and f_2 be two Boolean variables of $v_1(C)$ and $v_2(C)$ respectively that specify corresponding outputs of N_1 and N_2. Since S is a CS of N_1 and N_2, then $q_1(C) = q_2(C)$. So any assignment $q_1(c), q_2(c)$ to $v_1(C)$ and $v_2(C)$ that satisfies $Cf(v_1(C), v_2(C))$ also satisfies clauses $K'{=}f_1 \vee \sim f_2$ and $K''{=}{\sim}f_1 \vee f_2$. So K' and K'' are implied by $Cf(v_1(C),v_2(C))$ and can be deduced by the procedure described in the proof of Proposition 1.6. (The resolution steps one needs to deduce equivalence clauses are already counted in the expression $n * 3^{6p}$.)

Using each pair of equivalence clauses K' and K'' and the clauses specifying the gate g=XOR(f_1,f_2) of the miter, one can deduce a single literal clause $\sim g$. This clause requires setting the output of this XOR gate to 0. Each such a clause can be deduced in the number of resolutions bounded by a constant and the total number of such clauses cannot be more than $n*p$. Finally, from these unit clauses and the clauses specifying the final OR gate of the miter, the empty clause can be deduced in the number of resolutions bounded by $n*p$. So the empty clause is deduced in no more than $n*3^{6p} + d'*n*p$ steps where d' is a constant. Finally, one can pick a constant d such that $n*3^{6p} + d'*n*p \leq d*n*3^{6p}$

Remark 1.9 *In Proposition 1.7 we give a very conservative estimate of the complexity of deducing filtering and correlation functions. In practice this complexity can be much lower. In a sense, the best way to interpret the theory developed in this section is that the problem of equivalence checking of circuits N_1,N_2 with a CS S of n blocks can be partitioned into n subproblems of computing filtering and correlation functions for each variable of S.*

Remark 1.10 *In general, two functionally equivalent circuits N_1, N_2 may have more than one CS. In that case, when estimating the complexity of equivalence checking of N_1,N_2, it is natural to use the finest CS (see Definition 1.9).*

2.7 Equivalence Checking of Circuits with Unknown CS

In Section 2.6 we considered equivalence checking in general resolution that is a non-deterministic proof system. This means that the proof is guided by an "oracle" that points to the next pair of clauses to be resolved. Deterministic algorithms do not have the luxury of using an oracle. A natural question is whether a deterministic algorithm can benefit from the fact that the formulas from $M(p)$ have short proofs of unsatisfiability in general resolution. (In this section, we assume that one has to prove the unsatisfiability of a formula F, $F \in M(p)$ specifying equivalence checking of N_1,N_2 and no CS of N_1,N_2 is known.) A theory studying the complexity of finding proofs started only a few years ago [2, 18] and so it cannot fully answer this question yet. However, there is a good reason to believe that formulas of $M(p)$ are hard for deterministic algorithms. (Henceforth, by a deterministic algorithm we mean a resolution based deterministic SAT-algorithm.) Indeed, let us make the following two very plausible assumptions. First assumption is that there is a subclass M^* of formulas from $M(p)$ such that resolution proofs described in the proof of Proposition 1.7 (we will refer to them as ***specification driven proofs***) are "much shorter" than any other kind of resolution proofs. Second assumption is that finding a non-trivial CS of two Boolean circuits N_1 and N_2 is hard. If the two assumptions above are true then formulas from M^* should be hard. Indeed, specification driven resolution proofs very closely follow a CS of N_1 and N_2. So knowing a short resolution proof of the unsatisfiability of F,$F \in M^*$ one could easily recover the CS that "guided" that proof. That would mean that there is an efficient algorithm for extracting a common specification of N_1 and N_2, which contradicts our second assumption. One more argument in support of the conjecture that formulas from $M(p)$ are hard for deterministic algorithms is that formulas from $M(p)$ are hard for the best existing SAT-solvers (see Section 2.9).

To give the reader an idea of how big the difference between the size of non-deterministic and deterministic proofs might be, let us consider the class of formulas $M(p)$ where p is bounded by a constant. From Proposition 1.7 it follows that specification driven proofs consist of at most $d*n*3^{6p}$ resolution steps that is they have linear size. On the other hand, the complexity of these formulas for a deterministic algorithm should be $Length(F)^{g(p)}$ where F is a formula of $M(p)$, $Length(F)$ is the length of F and $g(p)$ is a monotone increasing function that is linear (or close to linear) in p. One argument in favor of such complexity is that a deterministic algorithm views the whole formula F as one "block" and the complexity of specification driven proofs is exponential in the size of the maximal block. Another reason is that as it was shown in [9] one can always pick binary encodings of multi-valued variables of a CS so that every specification driven proof will have to contain "long" clauses whose length is a monotone increasing function of p. Then even formulas from a class $M(p)$ with a quite small value of p, like p=10, can be extremely hard for a deterministic algorithm. So it is quite possible that no matter how good and efficient your resolution based SAT-solver is it will not be able to solve even formulas of linear complexity!

2.8 A Procedure of Equivalence Checking for Circuits with a Known CS

In the previous section, we gave some reasons why formulas from $M(p)$ should be hard for a deterministic resolution based SAT-algorithm. Let S be a CS of Boolean circuits N_1,N_2 and p be the granularity of S. Let F be the formula of $M(p)$ specifying the equivalence checking of N_1,N_2. The good news is that if S is known then there is an efficient algorithm for proving the unsatisfiability of F. This algorithm also proceeds in topological order of variables of S computing filtering and correlation functions. The only difference with specification guided proofs of general resolution is that the "power" of the proof "oracle" is limited. Namely, in general resolution this oracle guides every resolution step of the proof (pointing to the next pair of clauses to resolve). In the deterministic algorithm described below the specification S serves as an oracle of "limited" power. Namely, this oracle helps only to identify subcircuits $I_1(G)$ and $I_2(G)$ of N_1 and N_2 that are implementations of the same block $C = G(A, B)$. Finding the correlation function $Cf(v_1(C), v_2(C))$ and filtering functions $Ff(v_1(C))$ and $Ff(v_2(C))$ is done by this algorithm without any "help".

Our procedure of equivalence checking consists of two stages:

1. For each variable C of S compute filtering functions $Ff(v_1(C))$, $Ff(v_2(C))$ and the correlation function $Cf(v_1(C),\ v_2(C))$ proceeding in topological order of variables. If C is a primary input variable, then $Ff(v_1(C))$, $Ff(v_2(C))$ and $Cf(v_1(C),\ v_2(C))$ are tautologies. Let C=$G(A,B)$. Then $Ff(v_1(C))$ is built by computing the function existentially implied (see Definition 1.15) by $Ff(v_1(A)) \vee Ff(v_1(B)) \vee F(I_1(G))$. ($F(I_1(G))$ is a subset of F specifying the implementation of G in N_1. The function $Ff(v_2(C)$ is built similarly to $Ff(v_1(C))$.) The function $Cf(v_1(C)$,$v_2(C))$ is built by computing the function existentially implied by $Ff(v_1(A)) \vee Ff(v_1(B)) \vee Ff(v_2(A)) \vee Ff(v_2(B)) \vee Cf(v_1(A),\ v_2(A)) \vee Cf(v_1(B),\ v_2(B)) \vee F(I_1(G)) \vee F(I_2(G))$.

2. Once correlation functions are computed for all primary output variables of S, finish the proof of unsatisfiability of F by invoking a SAT-solver like [8],[16]. (This SAT-solver is applied to the CNF consisting of the clauses describing the correlation functions for the primary output variables of S, the clauses specifying the gates XORing primary outputs of N_1 and N_2 and the final OR gate of the miter.)

The complexity of this procedure is about the same as in general resolution which is equal to $d * n * 3^{6p}$ where d is a constant and n is the number of blocks. The only difference is that in general resolution no resolvent is generated twice while the procedure above may generate identical clauses when computing correlation or filtering functions. So it will have to take care of removing duplicate clauses.

The described procedure is flexible with respect to the method of computing existentially implied functions. Below we describe a few options. Let F be a CNF and $supp(F) = X_1 \cup X_2$. Suppose one needs to compute a CNF $H(X_2)$ that is existentially implied by F. If the value of $|X_2|$ is small, one can compute $H(X_2)$ by running 2^k SAT-checks where $k{=}|X_2|$. For every assignment z to the variables of X_2 one needs to check if there is an assignment y to the variables of X_1 such that (y,z) satisfies F. If such an assignment exists then the next assignment is checked. Otherwise, a clause consisting of literals of variables from X_2 that is falsified by the assignment z is added to the clauses of $H(X_2)$.

If the size of X_2 is large, one can compute filtering and correlation functions by existential quantification of the variables of X_1. In terms of SAT, existential quantification of a CNF F in a variable w of X_1 means adding to F all the resolvents that can be produced by resolving clauses of F in w. Of course, existential quantification in all the variables of X_1 is very expensive in SAT and so it works only for blocks of a small size. However, less expensive methods for computing $H(X_2)$ in terms of SAT can be and should be developed.

2.9 Experimental Results

The objective of experiments was to show that equivalence checking of circuits with a fine CS S is easy if S is known and is hard otherwise. To produce circuits having a fine CS we used the following procedure. To get multi-valued specifications with realistic topologies we "borrowed" them from MCNC-91 benchmark circuits as follows. First, all the benchmarks were technology mapped using SIS [20] consisting only of two-input AND gates. Then from each obtained circuit N a multi-valued specification S was produced by replacing each two-input binary gate with a two-input single output block of four-valued variables. (In other words, S changes the functionality of N while preserving its topology.) Then from S two functionally equivalent Boolean circuits N_1, N_2 implementing S were produced using two different sets of two-bit encodings of four-valued values. The encodings were picked in such a way that the two different implementations of the same four-valued block in N_1 and N_2 had no functionally equivalent outputs. This way we guaranteed that internal functionally equivalent points in N_1 and N_2 may occur only by accident.

Note that after encoding, the number of inputs and outputs in N_1 and N_2 is twice the number of inputs and outputs in the original Boolean circuit N. For instance, the two circuits produced from C6288 used as a "specification" have the topology of a 16-bit multiplier and the number of inputs and outputs of a 32-bit multiplier.

In experiments we used the best tools that were available to us. Namely, we used the SAT-solver BerkMin downloaded from [1], the program Nanotrav built on top of the Colorado University Decision Diagram (CUDD) package [6] and a SAT-based equivalence checker CSAT [14] (courtesy of Prof. Li of UCSB). We also tried the SAT-solver Zchaff [16], but BerkMin was up to three orders of magnitude faster on our formulas. In the experiments we used the special mode of BerkMin designed for equivalence checking that is described at [1]. BerkMin was run on the formula specifying the miter M of N_1 and N_2 as described in Section 2.3. Nanotrav was used to build a BDD for the miter M and CSAT checked the satisfiability of the miter's output. We first ran the three tools on "regular" MCNC benchmarks to verify optimized versus non-optimized circuits. (We do not report these results). The tools showed quite decent performance. For example, BerkMin was able to quickly verify all the instances including the multiplier C6288. The same kind of performance was shown by CSAT. Nanotrav was able to build BDDs for all the miters except C6288 very quickly (in a few seconds). In all the experiments we ran Nanotrav using settings suggested by Fabio Somenzi (private communication). In particular, the variable sifting option was on. In Table 2.9 we give runtimes of the three programs shown in our experiments. All the programs were run on a SUNW Ultra-80 system with clock frequency 450MHz. In all the experiments the time limit was set to 60,000 sec. (16.6 hours). The results of the best out of the three programs is shown in black. In the last column we report run times of a trivial CS driven procedure. This procedure computes filtering and correlation function of blocks in terms of SAT by existentially quantifying variables (as it was described in Section 2.8) and eventually deduces an empty clause.

It is not hard to see that run times of the CS driven procedure are linear in the size of circuits to be checked for equivalence. This is due to the fact that the size of specification blocks is fixed (and very small). On the other hand, the instances we generated turned out to be hard for the three chosen tools. Even if one compares the best run times with run times of the CS driven procedure, it is not hard to see that the former quickly increased as the size of the instances grew.

It is unlikely that an industrial strength equivalence checker would do much better on the circuits we generated because they have no functionally equivalent points. Besides, one can always produce much harder equivalence checking problems by using ***even a slightly*** more coarse specification (Recall that in the experiments we used a very fine CS

Table 1.1. Equivalence checking of circuits with a fine CS

Name of "specification"	Number of variables	Number of clauses	CSAT (sec.)	Nanotrav (BDDs) (sec.)	BerkMin (sec.)	CS driven (sec.)
C880	1,612	9,373	162.8	60,000	**3.7**	1.1
ttt2	2,770	17,337	281.0	**1.0**	11.7	1.3
x4	4,166	24,733	284.3	**4.7**	17.3	1.8
i9	4,954	29,861	75.3	**1.5**	32.7	2.1
term1	3,504	22,229	1,604.6	40.9	**35.9**	1.6
c7552	11,282	69,529	282.0	60,000	**52.8**	3.6
c3540	5,248	33,199	34,905.8	60,000	**64.1**	2.3
rot	5,980	35,229	163.6	19,315.6	**72.2**	2.1
9symml	960	6,105	31.07	**1.9**	113.2	0.5
frg2	10,316	62,943	13,610.4	**22.6**	131.4	2.9
frg1	3,230	20,575	**265.8**	60,000	330.3	1.7
i10	12,998	77,941	60,000	60,000	**445.0**	4.8
des	28,902	179,895	12,520.3	**9.7**	451.7	12.1
dalu	9,426	59,991	17,496.9	60,000	**518.6**	3.1
x1	8,760	55,571	13,580.3	13,009.6	**950.2**	2.8
alu4	4,736	30,465	8,020.4	**135.1**	992.6	2.0
i8	14,524	91,139	60,000	**98.0**	1,051.5	5.1
c6288	9,540	61,421	60,000	60,000	**1,955.1**	5.2
k2	11,680	74,581	60,000	59,392.9	**5,121.5**	4.3
too_large	58,054	376,801	60,000	60,000	60,000	15.2
t481	19,042	123,547	60,000	60,000	60,000	6.3

S consisting of four-valued blocks. That is the circuits produced from S were "almost" identical.) As we mentioned in the introduction, the problem of finding a short proof of equivalence of N_1,N_2 if a CS is not known, comes down to recovering this CS from the description of N_1,N_2 which is computationally very hard (if not infeasible).

2.10 Conclusions

In the first part of this chapter, we introduced a class $M(p)$ of CNF formulas specifying equivalence checking of Boolean circuits with a common specification (CS). We showed that formulas of $M(p)$ are "easy" for general resolution and gave reasons why those formulas should be hard for a deterministic algorithm that does not know a CS of the circuits to be checked for equivalence. We also gave some experimental evidence that formulas from $M(p)$ are hard for existing SAT-solvers. Besides, we formulated an efficient SAT-algorithm for equivalence checking of circuits with a known CS. The results of the first part of this chapter lead to the following two conclusions.

- A resolution based SAT-solver (most probably) cannot be scalable even on "easy" and practical formulas unless some extra information about the structure of short proofs is provided. (In case of equivalence checking this extra information is provided by a CS.)

- The SAT-solvers of the future should be very "intelligent" that is very receptive to structural properties of the formula to be tested for satisfiability.

3. Stable Sets of Points

3.1 Introduction

In the first part of this chapter, we showed that it is extremely important for a SAT-solver to be "receptive" to structural properties of CNF formulas. However, the existing algorithms are not very good at taking into account such properties. One of the reasons is that currently there is no "natural" way of traversing the search space. For example, in the DPLL procedure [7] which is the basis of almost all algorithms used in practice the search is organized as a binary tree. In reality, the search tree is used only to impose a linear order on the points of the Boolean space to avoid visiting the same point twice. However, this order may be in conflict with "natural" relationships between points of the Boolean space that are imposed by the CNF formula to be checked for satisfiability (for example, if this formula has some symmetries).

In the second part, we introduce the notion of a stable set of points (SSP) [11]. We believe that SSPs can serve as a basis for constructing algorithms that traverse the search space in a "natural" way. This may lead to creating SAT-solvers that are much more "intelligent" and efficient than the existing state-of-the-art SAT-solvers. We show that a CNF formula F is unsatisfiable if and only if there is a set of points of the Boolean space that is stable with respect to F. If F is satisfiable then any subset of points of the Boolean space is unstable, and an assignment satisfying F will be found in the process of constructing an SSP. We describe a simple algorithm for constructing an SSP. Interestingly, this algorithm is, in a sense, an extension of Papadimitriou's algorithm [17] (or a similar algorithm that is used in the well-known program called Walksat [19]).

A very important fact is that, generally speaking, a set of points that is stable with respect to a CNF formula F depends only on the clauses (i.e. disjunctions of literals) F consists of. So the process of constructing an SSP can be viewed as a "natural" way of traversing the search space when checking F for satisfiability. In particular, if F has symmetries, they can be easily taken into account when constructing an SSP. To illustrate this point, we consider the class of CNF formulas that are symmetric with respect to a group of permutations. We show that in this case for proving the unsatisfiability of a CNF formula it is sufficient to construct a set of points that is stable modulo symmetry.

If, for a class of formulas, SSPs are exponentially large, computing a monolithic SSP point-by-point is too time and memory consuming. We experimentally show that this is the case for hard random CNFs formulas. One of the possible solutions to this problem is to exclude some directions (i.e. variables) from consideration when computing an SSP. Such a set of points is stable only with respect to "movements" in the allowed directions. By excluding directions one can always get an SSP of small size. We sketch a procedure of satisfiability testing in which computing a monolithic SSP is replaced with constructing a sequence of small SSPs with excluded directions.

The second part of this chapter is structured as follows. In Section 3.2 we introduce the notion of an SSP. Section 3.3 relates an SSP with a set of points "reachable" from a point. A simple algorithm for building an SSP point-by-point is described in Section 3.4. We also show experimentally in Section 3.4 that even small CNF formulas may have large sets of SSPs and so computing SSPs point-by-point is in general infeasible. In Sections 3.5, 3.6 we discuss two possible ways of using SSPs. In Section 3.5 we show that to prove a symmetric CNF formula to be unsatisfiable it is sufficient to build an SSP modulo symmetries of that formula. Such an SSP can be sometimes efficiently built even point-by-point. Section 3.6 shows that the computation of a monolithic SSP can be replaced with the construction of so called SSPs with excluded directions whose size is easy to control. Finally, some conclusions are made in Section 3.7.

3.2 Stable Set of Points

In this section, we introduce the notion of an SSP. Let F be a CNF formula of n variables $x_1, \dots, x_n$. Denote by B the set $\{0, 1\}$ of values taken by a Boolean variable. Denote by B^n the set of points of the Boolean space specified by variables $x_1,\dots,x_n$. ***A point*** of B^n is an assignment of values to all n variables.

Definition 1.19 *Let p be a point of the Boolean space falsifying a clause C. The* ***1-neighborhood of the point*** *p with respect to the clause C (written* ***Nbhd(p,C)****) is the set of points that are at Hamming distance 1 from p and that satisfy C.*

Remark 1.11 *It is not hard to see that the number of points in Nbhd(p, C) is equal to that of literals in C.*

Example 1.4 *Let* $C = x_1 \vee \overline{x_3} \vee x_6$ *be a clause specified in the Boolean space of 6 variables* $x_1, \dots, x_6$. *Let* $p = (x_1 = 0, x_2 = 1, x_3 = 1, x_4 = 0, x_5 = 1, x_6 = 0)$ *be a point falsifying C. Then Nbhd(p, C) consists of the following three points:* $p_1 = (\boldsymbol{x_1{=}1}, x_2 = 1, x_3 = 1, x_4 = 0, x_5 = 1, x_6 = 0)$, $p_2 = (x_1 = 0, x_2 = 1, \boldsymbol{x_3{=}0}, x_4 = 0, x_5 = 1, x_6 = 0)$, $p_3 = (x_1 = 0, x_2 = 1, x_3 = 1, x_4 = 0, x_5 = 1, \boldsymbol{x_6{=}1})$. *Points* p_1, p_2, p_3 *are obtained from p by flipping the value of variables* x_1,x_3,x_6 *respectively i.e. the variables whose literals are in C.*

Denote by $Z(F)$ the set of points at which F takes value 0. If F is unsatisfiable, $Z(F) = B^n$.

Definition 1.20 *Let F be a CNF formula and P be a subset of Z(F). Mapping g of P to F is called a* ***transport function*** *if, for any* $p \in P$, *the clause* $g(p) \in F$ *is falsified by p. In other words, a transport function* $g{:}P \rightarrow F$ *is meant to assign each point* $p \in P$ *a clause that is falsified by p.*

Remark 1.12 *We call mapping* $P \rightarrow F$ *a transport function because, as it is shown in Section 3.3, such a mapping allows one to introduce some kind of "movement" of points in the Boolean space.*

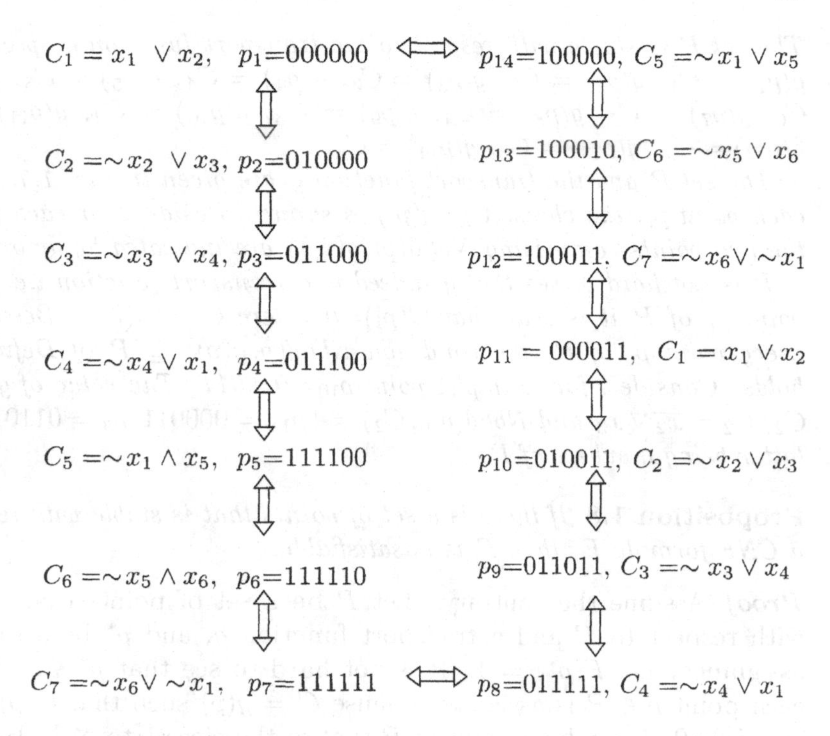

$C_1 = x_1 \vee x_2$, p_1=000000
$C_2 = \sim x_2 \vee x_3$, p_2=010000
$C_3 = \sim x_3 \vee x_4$, p_3=011000
$C_4 = \sim x_4 \vee x_1$, p_4=011100
$C_5 = \sim x_1 \wedge x_5$, p_5=111100
$C_6 = \sim x_5 \wedge x_6$, p_6=111110
$C_7 = \sim x_6 \vee \sim x_1$, p_7=111111
p_{14}=100000, $C_5 = \sim x_1 \vee x_5$
p_{13}=100010, $C_6 = \sim x_5 \vee x_6$
p_{12}=100011, $C_7 = \sim x_6 \vee \sim x_1$
p_{11} = 000011, $C_1 = x_1 \vee x_2$
p_{10}=010011, $C_2 = \sim x_2 \vee x_3$
p_9=011011, $C_3 = \sim x_3 \vee x_4$
p_8=011111, $C_4 = \sim x_4 \vee x_1$

Figure 1.7. Illustration to Example 1.5

Definition 1.21 *Let P be a nonempty subset of $Z(F)$, F be a CNF formula, and g: $P \rightarrow F$ be a transport function. The set P is called **stable** with respect to F and g if $\forall p \in P$, $Nbhd(p, g(p)) \subseteq P$. As it was mentioned before, "stable set of points" abbreviates to **SSP**.*

Remark 1.13 *Henceforth, if we say that a set of points P is stable with respect to a CNF formula F without mentioning a transport function, we mean that there is a function g:$P \rightarrow F$ such that P is stable with respect to F and g.*

Example 1.5 *Consider an unsatisfiable CNF formula F consisting of the following 7 clauses: $C_1 = x_1 \vee x_2$, $C_2 = \overline{x_2} \vee x_3$, $C_3 = \overline{x_3} \vee x_4$, $C_4 = \overline{x_4} \vee x_1$, $C_5 = \overline{x_1} \vee x_5$, $C_6 = \overline{x_5} \vee x_6$, $C_7 = \overline{x_6} \vee \overline{x_1}$. Clauses of F are composed of literals of 6 variables: $x_1,\ldots,x_6$. The following 14 points form an SSP P: p_1=000000, p_2=010000, p_3=011000, p_4=011100, p_5=111100, p_6=111110, p_7=111111, p_8=011111, p_9=011011, p_{10}= 010011, p_{11}=000011, p_{12}=100011, p_{13}=100010, p_{14}=100000. (Values of variables are specified in the order variables are numbered. For example, p_4 consists of assignments x_1=0, x_2=1, x_3=1, x_4=1, x_5=0, x_6=0.)*

The set P is stable with respect to the transport function g specified as: $g(p_1) = C_1$, $g(p_2) = C_2$, $g(p_3) = C_3$, $g(p_4) = C_4$, $g(p_5) = C_5$, $g(p_6) = C_6$, $g(p_7) = C_7$, $g(p_8) = C_4$, $g(p_9) = C_3$, $g(p_{10}) = C_2$, $g(p_{11}) = C_1$, $g(p_{12}) = C_7$, $g(p_{13}) = C_6$, $g(p_{14}) = C_5$.

The set P and the transport function g are given in Fig. 1.7. Next to each point p_i, the clause C_k=$g(p_i)$ is shown. Besides, for each point p_i the two points comprising Nbhd($p_i, g(p_i)$) are indicated by arrows.

It is not hard to see that g indeed is a transport function i.e. for any point p_i of P it is true that $C(p_i)$=0 where $C = g(p_i)$. Besides, for every point p_i of P, the condition Nbhd($p, g(p)$) $\subseteq P$ of Definition 5 holds. Consider, for example, point p_{10}=010011. The value of $g(p_{10})$ is C_2, $C_2 = \overline{x_2} \vee x_3$ and Nbhd(p_{10}, C_2) $= \{p_{11} = 000011, p_9 = 011011\}$, the latter being a subset of P.

Proposition 1.8 *If there is a set of points that is stable with respect to a CNF formula F, then F is unsatisfiable.*

Proof Assume the contrary. Let P be a set of points that is stable with respect to F and a transport function g, and p^* be a satisfying assignment i.e. $F(p^*) = 1$. It is not hard to see that $p^* \notin P$ because each point $p \in P$ is assigned a clause $C = g(p)$ such that $C(p)$=0 and so $F(p)$=0. Let p be a point of P that is the closest to p^* in Hamming distance. Denote by C the clause that is assigned to p by the transport function g i.e. $C = g(p)$. Denote by Y the set of variables values of which are different in p and p^*.

Let us show that C can not have literals of variables of Y. Assume the contrary, i.e. that C contains a literal of $x \in Y$. Then, since P is stable with respect to F and g, it has to contain the point p' which is obtained from p by flipping the value of x. But then $p' \in P$ is closer to p^* than p. So we have a contradiction. Since $C(p)$=0 and C does not contain literals of variables whose values are different in p and p^* we have to conclude that $C(p^*) = 0$. This means that p^* is not a satisfying assignment and so we have a contradiction.

Proposition 1.9 *Let F be an unsatisfiable CNF formula of n variables. Then set $Z(F)$ is stable with respect to F and any transport function $Z(F) \rightarrow F$.*

Proof Since F is unsatisfiable, then $Z(F) = B^n$. For each point $p \in B^n$, condition Nbhd($p, g(p)$) $\subseteq B^n$ holds.

Remark 1.14 *From propositions 1.8 and 1.9 it follows that a CNF F is unsatisfiable if and only if there is a set of points stable with respect to F.*

3.3 SSP as a reachable set of points

In this section, we introduce the notion of reachability that will be used in Section 3.4 to formulate an algorithm for constructing an SSP. Our main objective here is to show that the set of points reachable from a point of the Boolean space is an SSP unless this set contains a satisfying assignment.

Definition 1.22 *Let F be a CNF formula and g: $Z(F) \rightarrow F$ be a transport function. A sequence of k points $p_1, \ldots, p_k$, $k \geq 2$ is called a* **path** *from p_1 to p_k in a set P with a transport function g if points $p_1, \ldots, p_{k-1}$ are in P and $p_i \in Nbhd(p_{i-1}, g(p_{i-1}))$, $2 \leq i \leq k$. (Note that the last point of the path, i.e. p_k, does not have to be in P.) We will assume that no point appears twice (or more) in a path.*

Example 1.6 *Consider the CNF formula and transport function of Example 1.5. Let P be the set of points specified in Example 1.5. The sequence of points $p_1, p_{14}, p_{13}, p_{12}$ forms a path from p_1 to p_{12}. Indeed, it is not hard to check that $Nbhd(p_1, g(p_1)) = \{p_2, p_{14}\}$, $Nbhd(p_{14}, g(p_{14})) = \{p_{13}, p_1\}$, $Nbhd(p_{13}, g(p_{13})) = \{p_{14}, p_{12}\}$, $Nbhd(p_{12}, g(p_{12})) = \{p_{13}, p_{11}\}$. So each point p' of the path (except the starting point i.e. p_1) is contained in the set $Nbhd(p'', g(p''))$ where p'' is the preceding point.*

Definition 1.23 *Let F be a CNF formula. A point p'' is called* **reachable** *from a point p' by means of a transport function $g : Z(F) \rightarrow F$ if there is a path from p' to p'' with the transport function g. Denote by $Reachable(p, g)$ the set consisting of a point p and all the points that are reachable from p by means of the transport function g.*

Proposition 1.10 *Let F be a satisfiable CNF formula, p be a point of $Z(F)$, and s be a satisfying assignment (i.e. $s \notin Z(F)$) that is the closest to p in Hamming distance. Let g:$Z(F) \rightarrow F$ be a transport function. Then in $Z(F)$ there is a path from p to s with the transport function g i.e. the satisfying assignment s is reachable from p.*

Proof Denote by Y the set of variables whose values are different in p and s. Since $F(p)$=0, then $p \in Z(F)$ and the function g assigns a clause C to p where $C(p)$=0. All literals of C are set to 0 by p. On the other hand, since s is a satisfying assignment, then at least one literal of C is set to 1 by s. Then C contains a literal of a variable y from Y. Denote by p' the point obtained from p by flipping the value of y in p. The point p' is reachable from p by means of the transport function g. If $|Y| = 1$, then p' is the satisfying assignment s. If $|Y| > 1$, then p' cannot be a satisfying assignment since, by our assumption, the satisfying assignment s is the closest to p. Then after applying the same reasoning to the point p', we conclude that the clause assigned to p' by g must contain a literal of a variable y' from $Y \setminus \{y\}$. Flipping the value of y' in p' we produce a point p'' that is either the satisfying assignment s or is at distance $|Y| - 2$ from s. Going on in this manner we reach the satisfying assignment s in $|Y|$ steps.

Proposition 1.11 *Let P be a set of points that is stable with respect to a CNF formula F and a transport function $g : P \rightarrow F$. Then $\forall p \in P$, $Reachable(p, g) \subseteq P$.*

Proof Assume the contrary, i.e. that there is a point $p^* \in Reachable(p, g)$ that is not in P. Let H be a path from p to p^*. Denote by p'' the first point in the sequence of points specified by H that is not in P. (Points are numbered from p to p^*). Denote by p' the point preceding p'' in H. The point p' is in P and the latter is stable with respect to F and g. So $Nbhd(p', g(p')) \subseteq P$. The point p'' is in $Nbhd(p', g(p'))$ and so it has to be in P. We have a contradiction.

Proposition 1.12 *Let F be a CNF formula, $g : Z(F) \rightarrow F$ be a transport function, and p be a point from $Z(F)$. If $P = Reachable(p, g)$ does not contain a satisfying assignment for F, then P is stable with respect to F and g, and so F is unsatisfiable.*

Proof Assume the contrary i.e. that P is not stable. Then there exists a point p' of $Reachable(p,g)$ (and so reachable from p) such that a point p'' of $Nbhd(p',g(p'))$ is not in $Reachable(p,g)$. Since p'' is reachable from p' it is also reachable from p. We have a contradiction.

Remark 1.15 *From Proposition 1.12 it follows that a CNF F is satisfiable if and only if, given a point $p \in Z(F)$ and a transport function $g : Z(F) \rightarrow F$, the set $Reachable(p, g)$ contains a satisfying assignment.*

In [11] properties of SSPs are discussed in more detail.

3.4 Testing Satisfiability of CNF Formulas by SSP Construction

In this section, we describe a simple algorithm for constructing an SSP that is based on Proposition 1.12. Let F be a CNF formula to be checked for satisfiability. The idea is to pick a point p of the Boolean space and construct the set $Reachable(p, g)$. Since no transport function $g : Z(F) \rightarrow F$ is known beforehand, it is built on the fly. In the description of the algorithm given below, the set *Reachable*(p, g) is broken down into two parts: *Boundary* and *Body*. *Boundary* consists of those points of the current set *Reachable*(p, g) whose 1-neighborhood has not been explored yet. At each step of the algorithm a point p' of *Boundary* is extracted and a clause C falsified by p' is assigned as the value of $g(p')$. Then the set *Nbhd*(p', C) is generated and its points (minus those that are already in *Body*$\cup$ *Boundary*) are added to *Boundary*. This goes on until a stable set is constructed (F is unsatisfiable) or a satisfying assignment is found (F is satisfiable).

1 Generate a starting point p. *Boundary* $= \{p\}$. *Body*$=\varnothing$, $g = \varnothing$.

2 If *Boundary* is empty, then *Body* is an SSP and F is unsatisfiable. The algorithm terminates.

3 Pick a point $p' \in$ *Boundary*. *Boundary*=*Boundary* $\setminus \{p'\}$.

4 Find a set M of clauses that are falsified by point p'. If $M = \varnothing$, then the CNF formula F is satisfiable and p' is a satisfying assignment. The algorithm terminates.

5 Pick a clause C from M. Take C as the value of $g(p')$. Generate *Nbhd*(p', C). *Boundary* = *Boundary*$\cup$(*Nbhd*$(p', C)\setminus$*Body*). *Body* = *Body* $\cup \{p'\}$.

6 Go to step 2.

Interestingly, the algorithm described above can be viewed as an extension of Papadimitriou's algorithm [17] (or a similar algorithm used in the program Walksat [19]) to the case of unsatisfiable CNF formulas. Papadimitriou's algorithm (and Walksat) can be applied only to satisfiable CNF formulas since it does not store visited points of the Boolean space. An interesting fact is that the number of points that one has to explore to prove the unsatisfiability of a CNF formula can be very small. For instance, in example 1.5, an SSP of a CNF formula of 6 variables consists only of 14 points while the Boolean space of 6 variables consists of 64 points. It can be shown that for a subclass of the class of 2-CNF formulas (a clause of a 2-CNF formula contains at most 2 literals) the size of minimum SSPs grows linearly in the number of variables of the formula.

A natural question to ask is: "What is the size of SSPs for "hard" CNF formulas?". One example of such formulas are random CNFs for which general resolution was proven to have exponential complexity [5]. Table 1.2 gives the results of computing SSPs for CNF formulas from the "hard" domain (the number of clauses is 4.25 times the number of variables [15]). For computing SSPs we used the algorithm described above enhanced by the following heuristic. When picking a clause to be assigned to the current point p' of *Boundary* (Step 5), we give preference to the clause C (falsified by p') for which the maximum number of points of *Nbhd*(p', C) are already in *Body* or *Boundary*. In other words, when choosing the clause C to be assigned to p', we try to minimize the number of new points we have to add to *Boundary*.

We generated 10 random CNFs of each size (number of variables). The starting point was chosen randomly. Table 1.2 gives the average values of the SSP size and the share (percent) of the Boolean space taken by an SSP. It is not hard to see that the SSP size grows very quickly. So even for very small formulas it is very large. An interesting fact though is that the share of the Boolean space taken by the SSP constructed by the described algorithm steadily decreases as the number of variables grows.

Table 1.2. SSPs of "hard" random CNF formulas

number of variables	SSP size	#SSP/#All_Space (%)
10	430	41.97
11	827	40.39
12	1,491	36.41
13	2,714	33.13
14	4,931	30.10
15	8,639	26.36
16	16,200	24.72
17	30,381	23.18
18	56,836	21.68
19	103,428	19.73
20	195,220	18.62
21	392,510	18.72
22	736,329	17.55
23	1,370,890	16.34

The poor performance of the proposed algorithm on random CNF formulas suggests that computing a "monolithic" SSP point-by-point is too time and memory consuming. There are at least three ways of solving this problem. First way concerns computing SSPs for symmetric CNF formulas. In Section 3.5 we show that to prove that a symmetric CNF formula is unsatisfiable it suffices to build a set of points that is stable modulo symmetry. Such a set of points can be very small. Another way of dealing with the exponential blow-up of SSPs is described in Section 3.6. The idea is to exclude some directions (i.e. variables) from consideration when computing an SSP. This way the size of an SSP can be drastically reduced. By constructing an SSP with excluded directions one obtains a new implicate of the formula. By adding this implicate to the formula we make it "simpler" (in terms of the size of its SSPs). By computing SSPs with excluded directions and adding the corresponding implicates we replace the computation of a monolithic SSP with the construction of a sequence of small size SSPs. A third (and probably most promising) way of making SSP computation more efficient is to build SSP in big "chunks" clustering "similar" points. We do not study this idea here leaving it for future research.

3.5 Testing Satisfiability of Symmetric CNF Formulas by SSP Construction

In this section, we introduce the notion of a set of points that is stable modulo symmetry. This notion allows one to modify the algorithm of SSP construction given in Section 3.4 to take into account a formula's symmetry. The modification itself is described at the end of the section. We consider only the case of permutations. However, a similar approach can be applied to a more general class of symmetries e.g. to the case when a CNF formula is symmetric under permutations combined with the negation of some variables.

Definition 1.24 *Let $X = \{x_1, \ldots, x_n\}$ be a set of Boolean variables. A* ***permutation*** *π defined on set X is a bijective mapping of X onto itself.*

Let $F = \{C_1, \ldots, C_k\}$ be a CNF formula. Let $p = (x_1, \ldots, x_n)$ be a point of B^n. Denote by $\pi(p)$ the point $(\pi(x_1), \ldots, \pi(x_n))$. Denote by $\pi(C_i)$ the clause that is obtained from $C_i \in F$ by replacing variables $x_1, \ldots, x_n$ with variables $\pi(x_1), \ldots, \pi(x_n)$ respectively. Denote by $\pi(F)$ the CNF formula obtained from F by replacing each clause C_i with $\pi(C_i)$.

Definition 1.25 *A CNF formula F is called* ***symmetric*** *with respect to permutation π if the CNF formula $\pi(F)$ consists of the same clauses as F. In other words, F is symmetric with respect to π if each clause $\pi(C_i)$ of $\pi(F)$ is identical to a clause $C_k \in F$.*

Proposition 1.13 *Let p be a point of B^n and C be a clause falsified by p i.e. $C(p)=0$. Let π be a permutation of variables $\{x_1,\ldots,x_n\}$ and $C' = \pi(C)$ and $p' = \pi(p)$. Then $C'(p') = 0$.*

Proof Let $\delta(x_i)$ be the literal of a variable x_i that is present in C. This literal is set to 0 by the value of x_i in p. The variable x_i is mapped to $\pi(x_i)$ in the clause C' and the point p'. Then the value of $\pi(x_i)$ in the point p' is the same as that of x_i in p. So the value of literal $\delta(\pi(x_i))$ in the point p' is the same as the value of $\delta(x_i)$ in p i.e. 0. Hence, the clause C' is falsified by p'.

Remark 1.16 *From Proposition 1.13 it follows that if F is symmetric with respect to a permutation π then $F(p) = F(\pi(p))$. In other words, F takes the same value at points p and $\pi(p)$.*

The set of the permutations, with respect to which a CNF formula is symmetric, forms a group. Henceforth, we will denote this group by G. The fact that a permutation π is an element of G will be denoted by $\pi \in G$. Denote by 1 the identity element of G.

Definition 1.26 *Let B^n be the Boolean space specified by variables $X=\{x_1,\ldots,x_n\}$ and G be a group of permutations specified on X. Denote by **symm(p,p′,G)** the following binary relation between points of B^n. A pair of points (p,p') is in $symm(p,p',G)$ if and only if there is $\pi \in G$ such that $p' = \pi(p)$.*

Definition 1.27 *Points p and p' are called **symmetric** if they are in the same equivalence class of symm(p,p′,G).*

Definition 1.28 *Let F be a CNF formula that is symmetric with respect to a group of permutations G and P be a subset of $Z(F)$. The set P is called **stable modulo symmetry** with respect to F and a transport function g: $P \rightarrow F$ if for each $p \in P$, every point $p' \in Nbhd(p, g(p))$ is either in P or there is a point p'' of P that is symmetric to p'.*

Proposition 1.14 *Let B^n be the Boolean space specified by variables $X = \{x_1,\ldots,x_n\}$. Let p be a point of B^n, C be a clause falsified by p, and a point $q \in Nbhd(p,C)$ be obtained from p by flipping the value of a variable x_i. Let π be a permutation of variables from X, p' be equal to $\pi(p)$, C' be equal to $\pi(C)$, and $q' \in Nbhd(p',C')$ be obtained from p' by flipping the value of variable $\pi(x_i)$. Then $q' = \pi(q)$. In other words, for each point q of $Nbhd(p,C)$ there is a point q' of $Nbhd(p',C')$ that is symmetric to q.*

Proof The value of a variable x_k, $k \neq i$ in q is the same as in p. Besides, the value of the variable $\pi(x_k)$ in q' is the same as in p' (q' is obtained from p' by changing the value of the variable $\pi(x_i)$ and since $k \neq i$ then $\pi(x_k) \neq \pi(x_i)$). Since $p' = \pi(p)$, then the value of x_k in q is the same as the value of variable $\pi(x_k)$ in q'. On the other hand, the value of variable x_i in q is obtained by negation of the value of x_i in p. The value of the variable $\pi(x_i)$ in q' is obtained by the negation of the value of $\pi(x_i)$ in p'. Hence the values of the variable x_i in q and the variable $\pi(x_i)$ in q' are the same. So $q' = \pi(q)$.

Proposition 1.15 *Let F be a CNF formula, P be a subset of $Z(F)$, and $g : P \rightarrow F$ be a transport function. If P is stable modulo symmetry with respect to F and g, then the CNF formula F is unsatisfiable.*

Proof Denote by $K(p)$ the set of all points that are symmetric to the point p i.e. that are in the same equivalence class of the relation *symm* as p. Denote by $K(P)$ the union of the sets $K(p)$, $p \in P$. Extend the domain of transport function g from P to $K(P)$ in the following way. Suppose p' is a point that is in $K(P)$ but not in P. Then there is a point $p \in P$ that is symmetric to p' and so $p' = \pi(p)$, $\pi \in G$. We assign $C' = \pi(C)$, $C = g(p)$ as the value of g at p'. If there is more than one point of P that is symmetric to p', we pick any of them.

Now we show that $K(P)$ is stable with respect to F and g: $K(P) \rightarrow F$. Let p' be a point of $K(P)$. Then there is a point p of P that is symmetric to p' and so $p' = \pi(p)$. Then from Proposition 1.14 it follows that for any point q of *Nbhd*$(p, g(p))$ there is a point $q' \in$ *Nbhd*$(p', g(p'))$ such that $q' = \pi(q)$. On the other hand, since P is stable modulo symmetry, then for any point q of *Nbhd*$(p, g(p))$ there is a point $q'' \in P$ symmetric to q and so $q = \pi^*(q'')$, $\pi^* \in G$ (π^* may be equal to $1 \in G$ if q is in P). Then $q' = \pi(\pi^*(q''))$. Hence q' is symmetric to $q'' \in P$ and so $q' \in K(P)$. This means that *Nbhd*$(p', g(p')) \subseteq K(P)$ and so $K(P)$ is stable. Then according to Proposition 1.8, the CNF formula F is unsatisfiable.

Remark 1.17 *The idea of the proof was suggested to the author by Howard Wong-Toi [22].*

Proposition 1.16 *Let $P \subseteq B^n$ be a set of points that is stable with respect to a CNF formula F and transport function $g : P \rightarrow F$. Let P' be a subset of P such that for each point p of P that is not in P' there is a point $p' \in P'$ symmetric to p. Then P' is stable with respect to F and g modulo symmetry.*

Proof Let p' be a point of P'. Let q' be a point of *Nbhd*$(p',g(p'))$. Point p' is in P because $P' \subseteq P$. Since P is a stable set then $q' \in P$. From the definition of the set P' it follows that if q' is not in P' then there is a point $r' \in P'$ that is symmetric to q'. So each point q' of $Nbhd(p', g(p'))$ is either in P' or there is a point of P' that is symmetric to q'.

Definition 1.29 *Let F be a CNF formula, G be its group of permutations, p be a point of $Z(F)$, and g: $P \rightarrow F$ be a transport function. A set $Reachable(p, g, G)$ is called the set of points* ***reachable from p modulo symmetry*** *if a) the point p is in $Reachable(p, g, G)$ b) each point p' that is reachable from p by means of the transport function g is either in $Reachable(p, g, G)$ or there exists a point $p'' \in Reachable(p, g, G)$ that is symmetric to p'.*

Proposition 1.17 *Let F be a CNF formula, G be its group of permutations, p be a point of $Z(F)$, and $g : P \rightarrow F$ be a transport function. If the set P=$Reachable(p, g, G)$ does not contain a satisfying assignment, then it is stable modulo symmetry with respect to F and g and so F is unsatisfiable.*

Proof Assume the contrary, i.e. that P is not stable modulo symmetry. Then there is a point $p' \in P$ (reachable from p modulo symmetry) such that a point p'' of $Nbhd(p',g(p'))$ is not in P and P does not contain a point symmetric to p''. On the other hand, p'' is reachable from p' and so it is reachable from p modulo symmetry. We have a contradiction.

Remark 1.18 *From Proposition 1.17 it follows that a CNF F that is symmetric with respect to a group of permutations G is satisfiable if and only if, given a point $p \in Z(F)$, a transport function $g : Z(F) \rightarrow F$, the set $Reachable(p, g, G)$ contains a satisfying assignment.*

Let F be a CNF formula and G be its group of permutations. According to Proposition 1.17 when testing the satisfiability of F it is sufficient to construct a set $Reachable(p, g, G)$. This set can be built by the algorithm of Section 3.4 in which step 5 is modified in the following way. Before adding a point p'' from $Nbhd(p', C) \setminus (Body \cup Boundary)$ to $Boundary$ it is checked if there is a point p^* of $Boundary \cup Body$ that is symmetric to p''. If such a point exists, then p'' is not added to $Boundary$.

For highly symmetric formulas the difference between the SSPs and SSPs modulo symmetry can be huge. For example, for pigeon-hole formulas the size of SSPs is exponential in the number of holes while the size of minimum SSPS modulo symmetry is linear in the number of holes [11].

3.6 SSPs with Excluded Directions

Unfortunately, the theory developed in Section 3.5 does not help in solving CNF formulas that have no (or have very few) symmetries. In this section, we describe a different way of reducing the size of SSPs. The idea is to replace the computation of a single SSP with the construction of a sequence of SSPs whose stability is "limited". These SSPs are called SSPs with excluded directions. The key point is that by excluding some directions from consideration one can drastically reduce the size of SSPs. The construction of an SSP with excluded directions allows one to generate a new clause that is an implicate of the initial CNF formula. This clause can be added to the current formula, which makes the obtained formula simpler in terms of the size of SSPs. For the new formula we can again build an SSP with excluded directions deducing a new implicate of the formula. A sketch of the procedure of satisfiability testing based on constructing SSPs with excluded directions is given at the end of the section.

Definition 1.30 *Let F be a CNF formula.* ***A set of excluded directions*** *is a set E of literals that a) does not contain opposite literals of the same variable; b) there is no clause C of F such that all literals of C are in E.*

Definition 1.31 *Let F be a CNF formula and C be a clause of F. Let E be a set of excluded directions. Denote by* ***Nbhd(p,C,E)*** *the set of points of $Nbhd(p,C)$ that set to 1 only the literals of C that are not in E.*

Remark 1.19 *Since, according to Definition 1.30, there is at least one literal of C that is not in E, then $Nbhd(p,C,E)$ is nonempty.*

Example 1.7 *Let a point p be equal to $(x_1 = 0, x_2 = 0, x_3 = 0, x_4 = 1, x_5 = 1, x_6 = 1)$. Let a clause C of a CNF F be equal to $x_1 \vee x_3 \vee \overline{x_6}$ and the set E of excluded directions be equal to $\{x_4, \overline{x_6}\}$. The set $Nbhd(p,C)$ consists of points p_1, p_2 and p_3 obtained from p by flipping the values of variables x_1, x_3, x_6 respectively. On the other hand, set $Nbhd(p,C,E)$ consists only of points p_1, p_2 because the point p_3 sets to 1 an "excluded" literal, namely the literal $\overline{x_6}$ of E.*

Definition 1.32 *Let P be a nonempty subset of $Z(F)$, F be a CNF formula, and g: $P \rightarrow F$ be a transport function. Let E be a set of excluded directions. The set P is called* ***stable with respect to F, g***

and E *if a) each point p of P sets all the literals of E to 0; b) for each point p of P, Nbhd(p, g(p), E)* $\subseteq P$.

Proposition 1.18 *If there is a set of points that is stable with respect to a CNF formula F and a set E of excluded directions, then any assignment satisfying F has to set to 1 at least one literal of E. In other words, the clause obtained by the disjunction of the literals of E is an implicate of F.*

Proof Let P be a set of points that is stable with respect to F, a transport function g and a set E of excluded directions. Make the assignments setting all the literals of E to 0. Remove from F all the clauses that are satisfied by these assignments and remove from the rest of the clauses all the literals that are in E (since they are set to 0). The obtained formula F' is unsatisfiable because the set P is stable with respect to F' and a transport function g'. Indeed, according to Definition 1.31, each point p of P sets all the literals of E to 0. Then the clause $C = g(p)$ of F cannot be satisfied by the assignment setting a literal l of E to 0. (If a clause C is satisfied by this assignment, it must contain the literal $\bar{l}$ but then C cannot be falsified by p.) So all the clauses assigned to the points of P by g are still in F'. Denote by g' the transport function that maps a point p of P to the clause C' obtained from the clause $C = g(p)$ by removing all the literals of E. It is not hard to see that $Nbhd(p, C') = Nbhd(p, C, E)$. So for each point p of P it is true that $Nbhd(p, g'(p)) \subseteq P$.

Remark 1.20 *A set of points stable with respect to a CNF F and a set E of excluded directions can be constructed by the algorithm of Section 3.4 modified in the following way. At step 1 the algorithm generates a starting point setting all the literals from E to 0. At step 5 it generates set* $Nbhd(p', C, E)$ *instead of* $Nbhd(p', C)$.

Example 1.8 *Let* $p_1 = (x_1 = 0, x_2 = 0, x_3 = 0, x_4 = 0, x_5 = 0, x_6 = 0, x_7 = 0)$. *Let F be a CNF formula containing clauses* $C_1 = x_1 \vee x_2 \vee x_3, C_2 = \overline{x_1} \vee x_4 \vee x_5$ *(and maybe some other clauses). Let the set E of excluded directions be equal to* $\{x_2, x_3, x_4, x_5\}$. *Denote by* p_2 *the point obtained from* p_1 *by flipping the value of* x_1. *Taking into account that* p_1 *falsifies clause* C_1 *and* p_2 *falsifies clause* C_2 *we can form the following transport function g:* $g(p_1) = C_1, g(p_2) = C_2$. *It is not hard to see that the set of points* $P = \{p_1, p_2\}$ *is stable with respect to clauses* C_1, C_2, *transport function g, and set E. Indeed, since literals* x_2 *and* x_3 *of* C_1 *are in E then* $Nbhd(p_1, g(p_1), E) = \{p_2\} \subseteq P$. *On the other hand, since literals* x_4 *and* x_5 *of* C_2 *are in E then* $Nbhd(p_2, g(p_2), E) = \{p_1\} \subseteq P$. *From Proposition 1.18 we conclude that the clause* $C = x_2 \vee x_3 \vee x_4 \vee x_5$

equal to the disjunction of literals of E is an implicate of the formula F. On the other hand, it is not hard to see that C is actually the resolvent of clauses C_1 and C_2.

Remark 1.21 *From Example 1.8 it follows that for an unsatisfiable formula F we can always choose a set E of excluded directions so that there is a set of two points that is stable with respect to F and E. Indeed, due to completeness of general resolution, in F there is always a pair of clauses C_1 and C_2 that produce a new resolvent. Then we form the set E of excluded directions consisting of all the literals of C_1 and C_2 except the literals of the variable in which the two clauses are resolved.*

Below we sketch a procedure of satisfiability testing based on computing SSPs with excluded directions.

1 Compute an SSP P of a limited size trying to minimize the set E of excluded directions

2 Stop if a satisfying assignment is found. The formula is satisfiable.

3 Stop if $E = \varnothing$. The formula is unsatisfiable.

4 Add the deduced clause (disjunction of the literals of E) to the current CNF formula.

5 Go to step 1.

The idea of the procedure is that adding new implicates gradually reduces the complexity of the initial formula F in terms of the size of "monolithic" SSPs. The claim that the size of SSPs decreases is based on the following observations. Any set of points that is stable with respect to a CNF formula F is also stable with respect to a CNF $F \cup \{C\}$ where C is a clause. So by adding clauses we preserve the best SSPs seen so far and may produce even smaller ones. The latter follows from the fact that by adding new implicates we will eventually produce an empty clause (at step 3 of the procedure above) and any set of clauses containing an empty clause has an SSP consisting of only one point.

An important advantage of obtaining new implicates by computing SSPs with excluded directions is that directions can be excluded on the fly. The choice of directions to exclude should be aimed at the reduction of the size of the constructed SSP (that is the directions that may lead to the blow-up of the SSP should be excluded). Besides, when excluding directions one can make use of the information about the structure of the CNF formula to be tested for satisfiability.

3.7 Conclusions

In the second part of this chapter we show that satisfiability testing of a CNF formula reduces to constructing a stable set of points (SSP). An SSP of a CNF formula can be viewed as an inherent characteristic of this formula. We give a simple procedure for computing an SSP. As a practical application we show that the proposed procedure of SSP construction can be easily modified to take into account symmetry (with respect to variable permutation) of CNF formulas. Finally, we introduce the notion of an SSP with excluded direction and describe a procedure of satisfiability testing based on constructing such SSPs. We believe that developing the theory of SSPs may lead to creating SAT-algorithms that are much more efficient and "intelligent" than the ones implemented in the state-of-the-art SAT-solvers.

References

[1] BerkMin web page. http://eigold.tripod.com/BerkMin.html

[2] Bonet M.,Pitassi T., Raz R. *On interpolation and automatization for Frege systems.* SIAM Journal on Computing, 29(6):1939-1967, 2000.

[3] Brand D. *Verification of large synthesized designs.* Proceedings of ICCAD-1993, pp. 534-537.

[4] Bryant R. *Graph based algorithms for Boolean function manipulation.* IEEE Trans. on Computers, C(35):677-691.

[5] V.Chvatal, E.Szmeredi. *Many hard examples for resolution.* J. of the ACM,vol. 35, No 4, pp.759-568.

[6] CUDD web page. http://vlsi.colorado.edu/~fabio/

[7] M.Davis, G.Logemann, D.Loveland. *A Machine program for theorem proving.* Communications of the ACM, 1962,vol. 5,pp. 394-397.

[8] Goldberg E., Novikov Ya. *BerkMin: A fast and robust SAT-solver.* Design, Automation, and Test in Europe (DATE '02), pp. 142-149, March 2002.

[9] Goldberg E., Novikov Ya. *How good are current resolution based SAT-solvers.* presented at SAT-2003,Margherita Ligure - Portofino (Italy), May 5-8,2003.

[10] Goldberg E., Novikov Ya. *Equivalence Checking of Dissimilar Circuits.* Presented at IWLS-2003. Laguna Beach, California, USA,May 28-30,2003.

[11] E. Goldberg. *Testing Satisfiability of CNF Formulas by Computing a Stable Set of Points.* Proceedings of Conference on Automated Deduction, CADE 2002, pp.161-180.

[12] E. Goldberg. *Proving Unsatisfiability of CNFs locally.* Journal of Automated Reasoning. vol 28:417-434, 2002.

[13] A.Haken. *The intractability of resolution.* Theor. Comput. Sci. 39 (1985),297-308.

[14] F. Lu, L.-C. Wang, K.-T. Cheng, R. Huang. *A circuit SAT solver with signal correlation guided learning*, DATE-2003, pp. 892-898.

[15] D.Mitchell, B.Selman, H.J.Levesque. *Hard and easy distributions of SAT problems.* Proceedings AAAI-92, San Jose,CA, 459-465.

[16] M.W. Moskewicz, C.F. Madigan, Y. Zhao, L. Zhang, S. Malik. *Chaff: Engineering an efficient SAT-solver.* Proceedings of DAC-2001,pp. 530-535.

[17] C.Papadimitriou. *On selecting a satisfying truth assignment.* Proceedings of FOCS-91, pp. 163-169

[18] Razborov A., Alekhnovich M. *Resolution is not automatizable unless W[p] is tractable.* Proc. of the 42nd IEEE FOCS-2001, pages 210-219.

[19] B.Selman, H.Kautz, B.Cohen. *Noise strategies for improving local search.* Proceedings of AAAI-94,Vol. 1, pp. 337-343.

[20] E. Sentovich, K. Singh, C. Moon, H. Savoj, R. Brayton, A. Sangiovanni -Vincentelli, *Sequential circuit design using synthesis and optimization.* Proceedings of ICCAD, pp 328-333, October 1992.

[21] Silva J., Sakallah K. *GRASP: A Search Algorithm for Propositional Satisfiability.* IEEE Transactions of Computers, 1999, Vol. 48,pp. 506-521.

[22] H.Wong-Toi. *Private communication.*

[23] H.Zhang. SATO: *An efficient propositional prover.* Proceedings of CADE-1997, pp. 272-275.

Chapter 2

ADVANCEMENTS IN MIXED BDD AND SAT TECHNIQUES

Gianpiero Cabodi
Politecnico di Torino, Dip. di Automatica e Informatica, Turin, Italy
gianpiero.cabodi@polito.it

Stefano Quer
Politecnico di Torino, Dip. di Automatica e Informatica, Turin, Italy
stefano.quer@polito.it

Abstract This chapter covers mutual interactions between Boolean Satisfiability (SAT) solvers and Binary Decision Diagrams (BDDs). More precisely, the presentation is focused on approaches mixing methodologies, techniques, and ideas coming from both research domains. First of all, it gives some preliminary definitions and it presents the main differences and affinities between SAT and BDD manipulation algorithms. After that, it overviews some of the most notable efforts to integrate the two technologies either in a loose or in a tight way. It eventually provides some evaluations and hints for open problems and possible future work.

Keywords: Formal verification, model checking, Boolean satisfiability (SAT), binary decision diagrams (BDDs), reachability analysis

1. Introduction

Efficient algorithms to manipulate Boolean functions arising in real-world applications have become increasingly popular, over the last few years, in several areas of computer-aided design and verification. In this chapter we focus on two classes of such algorithms: Complete Boolean Satisfiability solvers, and symbolic manipulation of Binary Decision Diagrams.

R. Drechsler (ed.), Advanced Formal Verification, 45-76.

Given a propositional formula, the Boolean Satisfiability Problem (commonly abbreviated as *SAT*) consists of determining a variable assignment such that the formula evaluates to true, or establishing that no such assignment exists. While SAT solvers' pivotal role in theory complexity has been known for a while, they have more recently received growing research attention with the purpose of solving practical problems. Such problems include ATPG, formal verification, timing verification and routing of field-programmable gate arrays. As far as efficiency is concerned, although SAT is an NP-complete problem, or at least no polynomial algorithm to solve it is known, large practical instances have been worked out thanks to efficient implementation procedures [31, 19]. These procedures are based on elementary and extremely time-efficient steps, which consider one problem variable at a time and appropriately prune the overall search space. Furthermore, most such algorithms are now publicly available and easy to use once the problem has been modeled and coded in the proper format. However, since SAT do not work on a canonical representation, many sub-problem computations may get repeated, i.e., SAT-based techniques are potentially limited by time resources.

Binary Decision Diagrams (*BDDs*) are commonly used to implicitly represent large solution spaces in combinational and sequential problems that arise in synthesis and verification. A BDD is a directed acyclic graph constructed in such a way that its directed paths represent objects of interest (such as subsets, clauses, minterms, etc.). BDDs may achieve an exponential compression rate, as the number of vertices and edges (graph size) is often exponentially lower than the number of paths (from root to leaves). BDDs can be transformed by algorithms that visit all vertices and edges of the directed graph in some order. These algorithms take therefore polynomial time in the current size of the graph. Unfortunately, when new BDDs are created, some algorithms tend to significantly increase the number of vertices, potentially leading to exponential memory requirements. Similarly to what just described for SAT solvers, the order of elementary steps is critically important. To reduce this drawback, variable reordering techniques have been introduced. Variable order is usually chosen either statically, i.e., by pre-processing the input formula, or dynamically, i.e., by analyzing the outcome of previous steps. Nonetheless, even after almost two decades of intensive research in the area, BDDs have never been able to deal with the largest models and problem instances.

Keeping the previous considerations in mind, it is clear that SAT and BDD techniques are often presented as mutually exclusive alternatives. In general, a BDD approach is more suitable for capturing all solutions of the problem simultaneously. On the contrary, SAT decision trees have no variable ordering restrictions, and can therefore potentially manage larger problems. As a direct consequence, mixed approaches can potentially offer mechanisms for trading off space and time.

Some researchers have recently followed this path by addressing ways of making BDD and SAT tools interact and cooperate to the solution of a common task. In general this has been obtained in different application domains (e.g., general SAT, combinational circuit verification, Bounded Model Checking and Unbounded Model Checking, etc.), and exploiting various interaction schemes (e.g, master-slave interaction between BDD and SAT engines, BDD pre-processing, exploiting SAT techniques for symbolic reachability, etc.). We overview and classify some of the most promising works within this general framework.

More in particular, Section 2 introduces some basic concepts and the notation adopted in the rest of the chapter. Then, Section 3 reports some theoretical considerations regarding differences and similarities between SAT and BDD approaches. After that, Sections 4, 5 and 6 discuss some more practical methodologies selected among the most promising attempts to exploit the best from both methodologies. Finally, Section 7 concludes the chapter with some considerations and hints for open problems and possible future works.

2. Background

2.1 SAT Solvers

In this section we give a brief description of SAT-based tools. For a more detailed overview on SAT solvers and a complete list of references the reader should refer to the tutorial paper [39].

SAT solvers generally operate on problems for which a Boolean function is specified in *Conjunctive Normal Form* (*CNF*). This form is a two-level decomposition: The logical conjunction (AND) of one or more *clauses*, each of which consists of the logical disjunction (OR) of one or more *literals*. A *literal* is an instance of a variable or its complement. A *satisfying* assignment for a given CNF formula is thus a set of values for variables such that each individual clause is satisfied.

The most known complete algorithms for deciding satisfiability are based on the Davis-Putnam method [18] (*DP*), and variations of the Davis-Logemann-Loveland method [17] (*DLL*). The former approach, is based on *resolution* and performs existential elimination on the propositional variables. The procedure is repeated until the formula equals either 0 (unsatisfiable problem instance) or 1 (satisfiable problem instance). Resolution tends to be memory intensive as existential elimination often generates a large number of clauses. The latter approach, based on *backtrack search*, implicitly enumerates the space of possible binary assignments, looking for a satisfying one. A decision tree keeps track of current assignments and prunes the search by iteratively applying *unit propagation*, usually referred to as *Boolean Constraint Propagation* (*BCP*). If a conflict is reached, the search backtracks to some previous assignment. *Conflict analysis*, and *recursive learning* constitute major enhancements to the basic backtrack search procedure.

- Conflict analysis comes into play when a conflict arises. It adds adequate information, a conflict clause, to anticipate the possible re-occurrence of the same conflict. Furthermore, conflict analysis allows the search process to backtrack *non-chronologically* to earlier levels in the search tree, considerably pruning the search space.
- Recursive learning, when extended to conjunctive normal form (CNF) clauses, identifies necessary assignments by examining the different possible ways of satisfying a given clause from the set of unassigned literals.

2.2 Binary Decision Diagrams

Binary Decision Diagrams (*BDDs*) [9] are directed acyclic graphs providing a canonical representation of Boolean functions. An Ordered BDD (*OBDD*) is a tree-like graph where Shannon (or Boole) decomposition $f = v \wedge f|_{v=1} \vee \overline{v} \wedge f|_{v=0}$ is recursively applied at each node, following an ordered set of variables. A Reduced OBDD (*ROBDD* [9]) for a given Boolean function is obtained by repeatedly applying two well known reduction rules:

- *Merging*, i.e., two isomorphic subgraphs are merged. The rule guarantees keeping a unique (canonical) representation (and BDD subgraph) for any given sub-function.
- *Deletion*, i.e., a BDD node whose two outgoing edges point to the same successor is deleted (see Figure 2.1). The rule represents the fact that a given sub-function does not depend on the deleted variable.

Notice that the term BDD is often used to "informally" denote a ROBDD, or as a more generic term indicating one of the several decomposition types proposed, as variants of the original ROBDDs. We will also use *DD* to indicate "generic" Decision Diagrams.

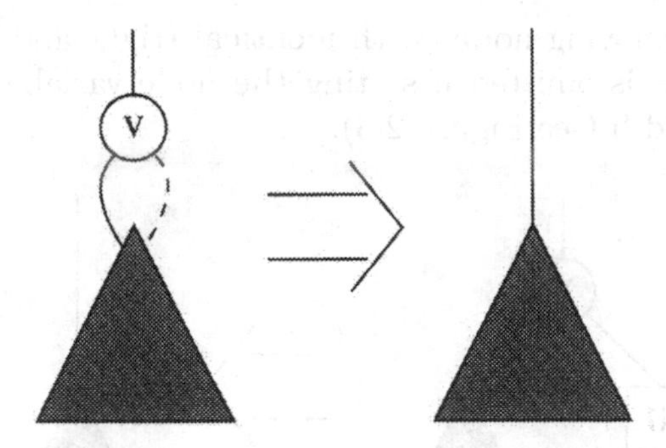

Figure 2.1. BDD Deletion Rule. A BDD node with two equal outgoing edges is deleted.

Example 2.1 Figure 2.2 shows the BDD for the function

$$f(x_1, x_2, x_3) = (x_1 \wedge \overline{x_2} \wedge x_3) \vee (\overline{x_1} \wedge x_2 \wedge x_3)$$

A solid (dashed) line indicates the 1 (0) value for the decision variable.

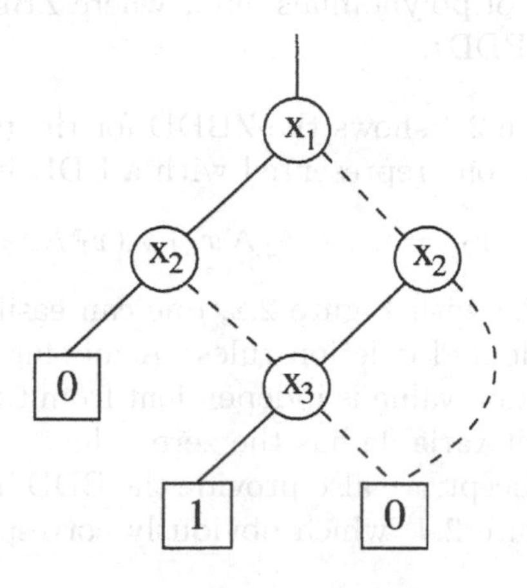

Figure 2.2. An Example of BDD.

Simple graph algorithms, working *depth–first* on BDDs, implement many operators. APPLY, ITE (if–then–else), and existential/universal quantifiers are well–known examples. BDDs have been widely used in verification problems to represent functions, as well as sets, by means of their characteristic functions. Operations on sets are efficiently implemented by Boolean operations on their characteristic functions.

2.2.1 Zero-Suppressed Binary Decision Diagrams.

Zero-Suppressed Binary Decision Diagrams (*ZBDDs*) [30] are a variant of BDDs adopting an alternative deletion rule.

Instead of removing nodes with identical (right and left) children, in a ZBDD a node is omitted if setting the node variable to 1 causes the function to yield 0 (see Figure 2.3).

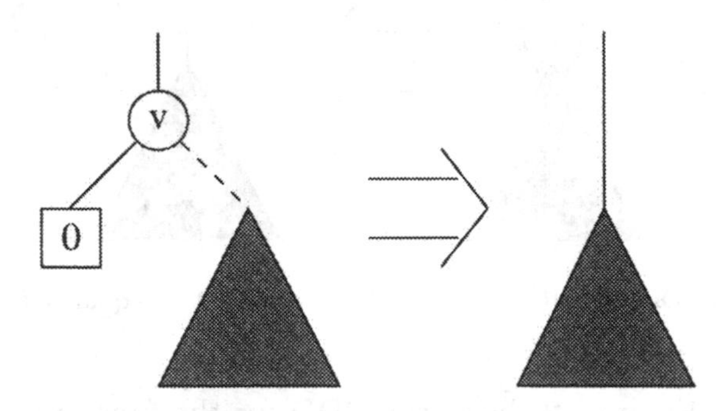

Figure 2.3. Reduction rules for ZBDDs.

The above rule is quite effective with sparse set representations, implicit manipulations of polynomials, etc., where ZBDDs are more compact than standard BDDs.

Example 2.2 Figure 2.4 shows the ZBDD for the function used in the Example 2.1, i.e., the one represented with a BDD in Figure 2.2:

$$f(x_1, x_2, x_3) = (x_1 \wedge \overline{x_2} \wedge x_3) \vee (\overline{x_1} \wedge x_2 \wedge x_3)$$

Comparing Figure 2.4 with Figure 2.2, one can easily notice the difference between the adopted deletion rules. A missing variable in a BDD means that the function value is independent from the variable, while in a ZBDDs the implicit variable has the zero value.

To enforce this concept, we also provide the BDD interpretation of the ZBDD graph of Figure 2.4, which obviously corresponds to a different Boolean function:

$$f(x_1, x_2, x_3) = (x_1 \wedge x_3) \vee (\overline{x_1} \wedge x_2 \wedge x_3)$$

2.2.2 Boolean Expression Diagrams. Boolean Expression Diagrams (*BEDs*) [24] are another extension of BDDs. They allow not only terminal and variable vertices, but also operator vertices. Terminal vertices correspond to the 0 and 1 constant leaves as in BDDs. Variable vertices v have the same semantics as vertices in a BDD, i.e., they correspond to the *if–then–else* operator of Boole's decomposition $((v \wedge f|_{v=1}) \vee (\overline{v} \wedge f|_{v=0}))$. Operator vertices *op* correspond to their respective Boolean connectives, i.e., $f|_{left}\, op\, f|_{right}$

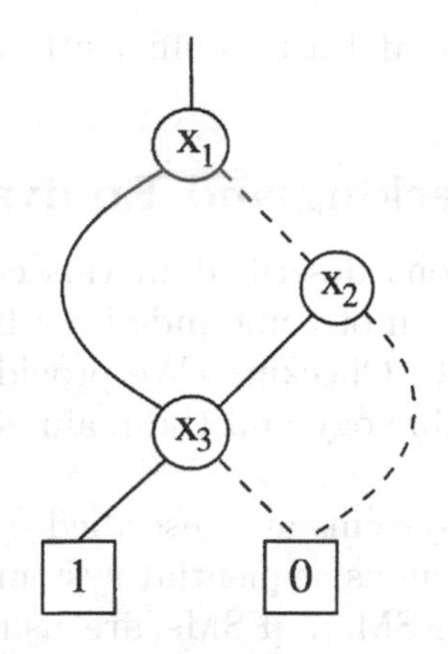

Figure 2.4. An Example of ZBDD.

Following this description a BDD is simply a BED without operator nodes, whereas a circuit can be directly mapped to a BED with operator nodes at all gates but primary inputs.

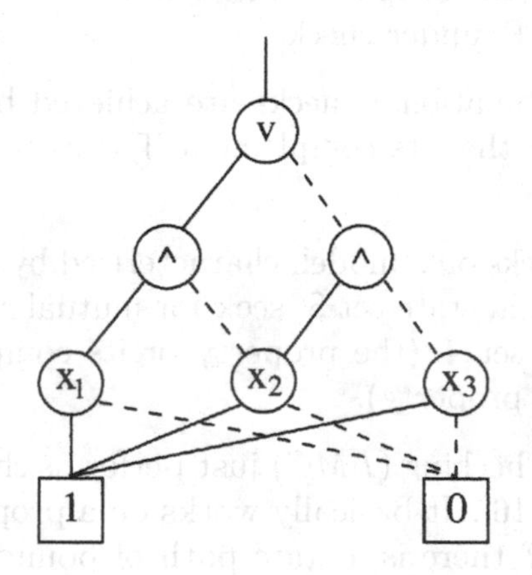

Figure 2.5. BED Example.

Example 2.3 Figure 2.5 shows a BED for function

$$f(x_1, x_2, x_3) = (x_1 \wedge x_2) \vee (x_2 \wedge x_3)$$

and its "corresponding" circuit.

BEDs can represent any Boolean formula in linear space at the price of being non-canonical. However, many of the desirable properties of BDDs are maintained, since converting a Boolean formula into a BDD via a

BED can always be done at least as efficiently as directly constructing the BDD.

2.3 Model Checking and Equivalence Checking

Most of the applications described in this chapter target hardware verification, under the form of Unbounded (or Bounded) Model Checking [10, 28] or Equivalence Checking. We provide in this section a brief introduction to the terminology and the main issues in BDD and SAT-based Model Checking.

While combinational systems are described by the Boolean functions of their outputs, synchronous sequential systems are often modeled as Finite State Machines (FSMs). FSMs are usually described by their Transition Relation TR, representing the present-next state behavior, and an initial set of states S.

Properties are expressed in terms of Boolean formulas (invariants) or temporal formulas (e.g., CTL properties). Equivalence checks are usually brought to invariant checks through the so called *miter* (or *product machine*) structure, where two circuits share their primary inputs, and extra gates compare the outputs under check.

Given a property P under check

- Standard combinational checks are achieved by proving that P is a tautology, or that its complement T (target state set, $\mathsf{T} = \overline{\mathsf{P}}$) is unsatisfiable.

- Sequential checks on a model, characterized by a transition relation TR and an initial state set S, seek for mutual reachability of S and a target state set T (the property or its complement, depending on the kind of property).

Bounded Model Checking (*BMC*) just performs checks up to a limited sequential depth [7, 16]. It basically works on a propositional formula f that is satisfiable *iff* there is a state path of bounded length k from S to T. Let us use s_i for the state or the $i-th$ step, with $1 \leq i < k$ (see Figure 2.6), then a BMC problem can be expressed as follows:

$$f = \mathsf{TR}(\mathsf{S}, s_1) \wedge \ldots \wedge \mathsf{TR}(s_{k-1}, s_k) \wedge (s_k = \mathsf{T})$$

Due to the above mentioned bounded depth, this is an incomplete verification technique, and it works well for falsification and partial verification. If correctness, rather than bug hunting, is the target, full verification is usually attempted by BMC with longer and longer bounds, possibly augmented with fix-point checks or inductive proofs. BMC is usually operated by SAT solvers.

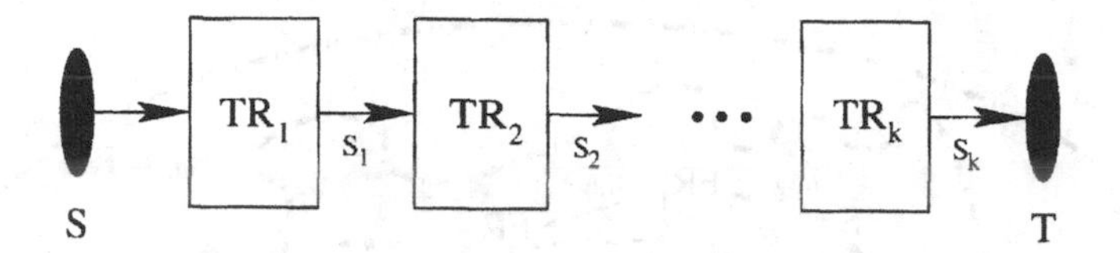

Figure 2.6. BMC propositional formula construction.

As far as BDDs are concerned, sequential verification is based on a symbolic traversal, i.e., a forward (or backward) breadth-first visit of the state space, represented by the following least fix-point (*lfp*) iteration:

$$\mathsf{FR} = lfp\ \mathsf{R}.(\mathsf{S} \vee (\textsc{Img}(\mathsf{TR}, \mathsf{R}))$$

The resulting state set FR is the set of forward reachable states. The method is based on the iterated application of the IMG operator, computing symbolic images of the R state set. Figure 2.8 shows a diagrammatic representation of the state space during the overall methodology.

As T may be reached before the fix-point, it is possible to avoid full computation of FR with on the fly tests for intersection with T. This is shown in Figure 2.7.

```
SEQUENTIALVERIFICATION (TR, S, T)
    New ← S
    From ← S
    FR ← S
    while (New ≠ ∅)
        if ((New ∧ T) ≠ ∅)
            return (Satisfiable)
        To ← IMG (TR, From)
        New ← To ∧ ¬FR
        FR ← FR ∨ New
        From ← BESTBDD (New, FR)
    return (Unsatisfiable)
```

Figure 2.7. Symbolic Forward Verification.

CTL model checking procedures are often implemented as backward traversals. This is easily expressed by swapping the S and T sets, and changing the IMG function with PREIMG.

Approximate Traversals [14, 20] are a possible way to extend the applicability of reachability analysis (with partial verification by sufficient checks) to larger circuits. The approach is based on the *approximate image* ($\textsc{Img}^+$) operator, returning over-estimations of exact images. Efficiency comes from the fact that, although R^+ represents more states

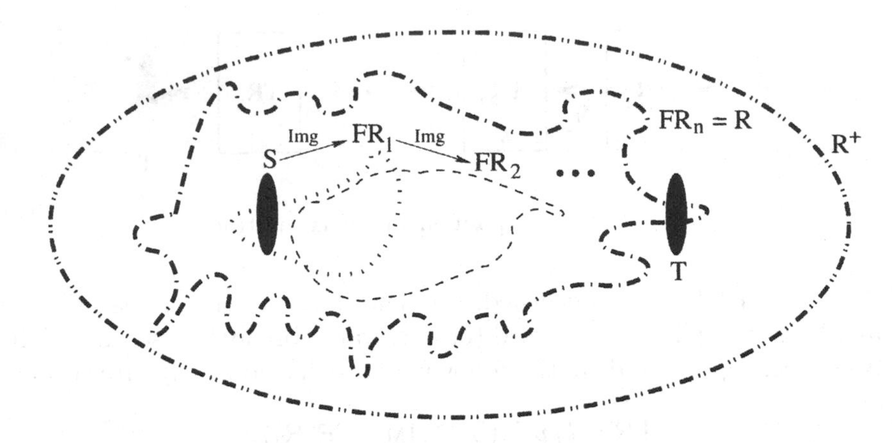

Figure 2.8. Reachability Analysis: Image Computation, R, and R^+ representation.

(see Figure 2.8), its BDD is usually much smaller than R (for that reason R^+ is represented with a smoother line with respect to R in Figure 2.8) as many mutual interactions and dependencies among state variables are artificially ignored.

3. Comparing SAT and BDD Approaches: Are they different?

Several comparisons between SAT and BDD based approaches have been attempted over the last few years, both from a theoretical and an experimental points of view. Some works are mainly focused on differences, and they show that the two methodologies have different classes of tractable instances. Other ones follow the intuition that common subproblems (e.g., variable ordering) could be faced by mixed techniques and heuristics.

In this section we overview some works in both of the above mentioned categories.

3.1 Theoretical Considerations

Most of the research effort showing differences between the SAT and BDD worlds is based on finding examples of different performance, either on specific benchmarks or broader industrial examples.

Uribe *et al.* [36] first show examples of out-performance in both directions between the two approaches. Using classical benchmarks from constraint satisfaction, artificial intelligence and combinational equivalence verification, they compare the original DP procedure with BDDs. They prove (theoretically as well as experimentally) that either technique may be exponential, while the other one is polynomial. They conclude that, in general, BDDs are well suited for representing large numbers of solutions that share a recursive structure, and for functional equivalence problems from the circuit verification domain. The favorable best performance of BDDs often comes at the expense of exponential memory usage. BDDs are found impractical to solve highly constrained problems with few or no solutions, such as the quasi-group and hard random 3-SAT problems, on which the DP procedure is good at. The DP procedure is also clearly superior if one only wants to find a single solution in a space rich of solutions. On the other hand, it practically was of no use in proving Boolean equivalences for which BDDs are particularly adept.

An interesting observation of [36] is that, when solving constraint satisfaction problems with BDDs, there basically is no search. To this respect the process consists entirely of constructing a BDD representation for a Boolean function that satisfies a given set of constraints. Consequently, the final result represents all possible solutions. A BDD based approach should thus be more fairly compared with exhaustive search algorithms, and not with those methods that find only one solution. However, this distinction is irrelevant in all cases where there are no solutions at all, since any procedure has to "exhaust" the space of possible solutions in one way or another.

More recently, Groote *et al.* [21] came to similar theoretical conclusions, i.e., "resolution-based and BDD-based approaches cannot simulate each others polynomially". They prove this property by showing formulas that can be solved polynomially with resolution and exponentially with BDDs or vice-versa.

Even though quite interesting, especially from a theoretical point of view, the above works should be just considered as initial attempts to compare the two approaches. Being focused on very limited sets of benchmarks, they only provide specific impressions, with some lack of generality, and they say very little about the real relation between resolution and BDDs. Experimental results on selected benchmarks may be influenced by badly chosen variable orders in BDDs or non optimal proof search strategies in SAT. Furthermore, new advancements in either technique (as for instance the dramatic recent improvements of SAT solvers in circuit verification), may even fully revise any previous statement or conclusion.

3.2 Experimental Benchmarking

A much more extensive experimental benchmarking has been proposed for the specific case of Bounded Model Checking (see Section 2.3).

The works described in this section [16, 8, 11] concentrate both on specific home-made benchmarks and on industrial examples.

3.2.1 Bug Hunting in an Industrial Setting.

Bjesse *et al.* [8] specifically target hardware verification to find bugs in a memory sub-system of the Alpha microprocessor.

As the initial sub-system had something like 14400 latches, 400 primary inputs, and 15 stages of pipeline, authors had to reduce its size before verification. To obtain a proper reduction, they applied different methodologies. In particular, as they do not need to preserve all possible properties as long as they can still find problems in the reduced circuit which are also present in the full size circuit, they also applied *formally incorrect* ad-hoc reductions. The final model has something like 600 latches.

On the resulting circuit, authors apply BDD-based symbolic model checking, SAT-based BMC, and Symbolic Trajectory Evaluation (*STE*).

Their experimental analysis shows that symbolic model checkers have a capacity limit which prevents "cost effective" bug hunting. For example, the BDD-based symbolic model checker CADENCE SMV needed from several hours to days to check simple properties. As a consequence, authors looked into SAT-based BMC strategies, using both the FIXIT tool and the publicly available SAT-solver GRASP [27]. Figure 2.9 shows a diagrammatic representation of the results by comparing the various CPU times as a function of the bound, i.e., the length of the failure checked.

As far as STE is concerned, authors discover that it can potentially allow much deeper explorations (preserving acceptable run times), but it requires much more time to produce good specifications. Their final conclusion is that the three methods have very different characteristics, so they finally propose a mixed verification methodology. Their suggestion is to start verification by SAT-based BMC with small bound. For each bug found, remove the bug, correct the model and move to larger bounds. Start working contextually with SAT-based BMC and STE whenever BMC takes more than half an hour. If neither BMC nor STE are able to find a failure, try SMV to complete the verification task.

3.2.2 Modifying BDD-based Techniques to Perform BMC.

Copty *et al.* [16] compare two internal Intel tools performing BMC: THUNDER and FORECAST. THUNDER is based on the DLL SAT solver SIMO. FORECAST, uses a traditional BDD package. The BDD package is an internal Intel tool, allowing dynamic reordering, partitioned tran-

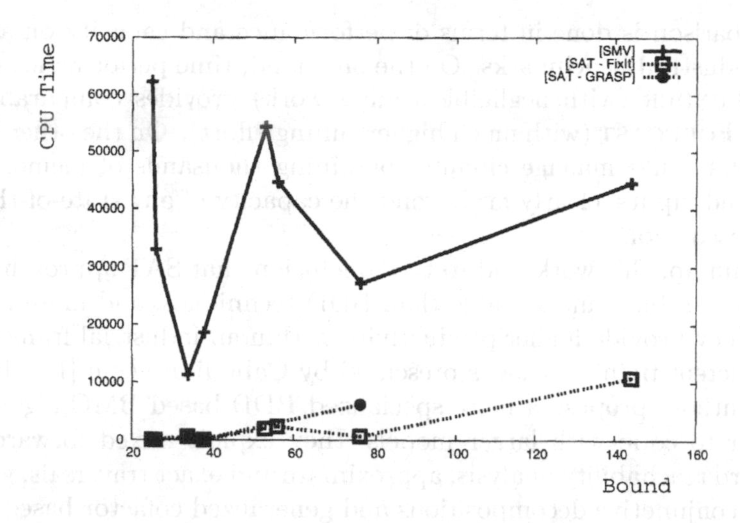

Figure 2.9. BDD and SAT comparison.

sition relations, prioritized traversal, and other state-of-the-art features. As BDD-based tools usually perform unbounded checks, authors transformed it to perform bounded verification. To this respect, as termination, i.e., checking for the fix-point, is not an issue in bounded search, authors modify the algorithm presented in Figure 2.7 as described in Figure 2.10.

```
SEQUENTIALBOUNDEDVERIFICATION (TR, S, T, k)
    Frontier ← S
    for (i ← 0; i ≤ k; i ← i + 1)
        if ((Frontier ∧ T) ≠ ∅)
            return (Satisfiable)
        Frontier ← IMG (TR, Frontier)
    return (Unsatisfiable)
```

Figure 2.10. Symbolic Forward Bounded Verification.

In this version of the algorithm the overall reached set of states, Reached, is not maintained. At the same time, to have manageable size for the Frontier state sets, authors adopt a *partitioned-prioritized* traversal. Basically, the Frontier set is partitioned whenever its BDD gets larger than a certain threshold and partitions are maintained in a priority queue. Partitions are inserted in the priority queue according to their distance from S, and they are discarded whenever they reach the maximum allowed bound.

Comparison is done in terms of performance and capacity on a wide set of industrial benchmarks. On the one hand, time performance shows that THUNDER (with negligible tuning work) provides comparable results as FORECAST (with much higher tuning effort). On the other hand, THUNDER could manage circuits containing thousands of memory elements and inputs, clearly far beyond the capacity of any state-of-the-art BDD-based tool.

To sum up, this work lead to the conclusion that SAT approaches are both more robust and scalable than BDD techniques, and more importantly they provide higher productivity within an industrial framework.

A different point of view is presented by Cabodi *et al.* in [11]. In this work, authors propose a more specialized BDD-based BMC algorithm. In order to cope with larger models, they exploit mixed forward and backward reachability analysis, approximate and exact traversals, guided search, conjunctive decompositions and generalized cofactor based BDD simplifications. The overall method could attack models in the range up to 1000-2000 memory elements. To this respect, BDDs seem to be more scalable with increasing bounds, while SAT tools are able to attack bugs on larger problems. Figure 2.11 shows this behavior on a well known benchmark, while checking an invariant property available with the original circuit description.

The graph compares CPU times for the BDD-based tool (called FBV) and the SAT-based BMC tool (NuSMV [15], version 2.0.2), with increasing value of the bound. The SAT-based tool uses both the internal and a state-of-the-art SAT solver (mchaff [31]) as slave engines.

The authors argue that the BDD based approach can be considered as complementary to SAT-based BMC, especially when looking for sequentially deep bugs.

3.2.3 Conclusions. To sum up, the previous works show that SAT-based tools are generally more robust and scalable than BDD techniques. At the same time it is also true that there seems to be enough room also for BDD-based techniques especially as far as hard corner cases and ad-hoc techniques are concerned. In the following sections we concentrate on affinities and collaborative techniques.

3.3 Working on Affinities: Variable Order

On a completely orthogonal direction, the emphasis is put in [2, 3, 4, 34] on the affinities between BDDs and SAT. More specifically, the works concentrate on strategies to find a good variable order or to select a good variable when performing a specific step of the algorithm.

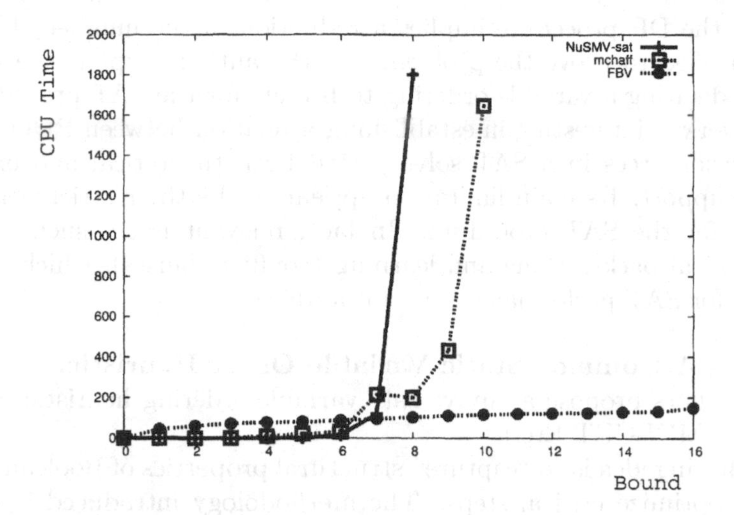

Figure 2.11. BMC execution time (sec) versus bound value for the am2901 benchmark.

It is well known that a proper variable order can provide an exponential compression rate. A similar behavior is experienced with SAT solvers. GRASP [27], for example, uses dynamic variable-ordering heuristics, such as DLCS (Dynamic variation of LCS) or DLIS (Dynamic variation of LIS). The former selects the variable that appears in the maximum number of unresolved clauses, while the latter one selects the literal with the same property.

3.3.1 Affinities on circuit-width correlation. Affinities between SAT and BDDs were already pointed out by Prasad *et al.* [33]. Their work mainly studies circuit SAT for ATPG, with the aim of characterizing its complexity in terms of circuit cut-width. As a byproduct, authors argue that BDDs show a similar behavior, since circuit-width is an upper bound to their size.

3.3.2 Recursion tree and Variable Order. Reda *et al.* [34] study the relation between the search tree of the DP/DLL procedures and the BDD of the corresponding function. They establish that the number of paths from the root node to the terminals of the BDD is directly related to the number of backtracks needed to prove the equivalence of two functionally equivalent circuits. This relation introduces the ability to calculate an optimal lower bound on the number of backtracks needed to prove equivalence. In addition, this relation leads to the conclusion that the capture of the variable ordering of the minimal path

BDD in the DP procedure implies a reduction in the number of backtracks needed to prove the problem. So the authors exploit the above fact, by devising a variable ordering technique for the SAT procedure.

The work is interesting in establishing a relation between BDD paths and decision trees in a SAT solver, with both theoretical and experimental support. Its main limitation appears to be the restricted model adopted for the SAT procedure. In fact, relevant issues such as non chronological backtracking and learning (conflict clauses), which play a key role for SAT performance, are not addressed.

3.3.3 A Common Static Variable Order Heuristic. In [3, 2] the authors propose a "universal" variable-ordering heuristic, called MINCE (MIN CUT Etc.).

The driving idea is to "capture" structural properties of Boolean functions to optimize critical steps. The methodology introduced is based on a variable partitioning (and consequently reordering) scheme able to capture "connections" among variables. If one is able to partition all variables into two largely independent separate groups, then the function is likely to be represented by a BDD with small cut, i.e., with few edges between the two groups of variables. BDDs with small cuts tend to have fewer edges and vertices.

To obtain this, the authors build an hyper-graph, whose vertices correspond to variables and edges to clauses of the original CNF formulation of the problem. Then, they recursively partition the graph by using the CAPO tool, applying min-cut partitioning with several optimizations. In this way, they are able to isolate unrelated or loosely related subsets of variables. The linear placement generated by the partitioning heuristic is then translated back into an ordering of CNF variables and used both for the SAT and the BDD tools.

Authors claim benefit in using the MINCE variable ordering both to drive SAT decisions and as a static variable ordering for BDDs.

3.3.4 Conclusions. Although attractive in their starting goal, the above works just cover a very particular aspect of BDD and SAT interactions. Furthermore, the experimental support is still rather preliminary and do not always keep into consideration state-of-the-art tools or recent improvements in the tools and heuristics.

4. Decision Diagrams as a Slave Engine in general SAT: Clause Compression by Means of ZBDDs

Let us move now to examples of mixed approaches. We describe in this section some techniques, derived from typical SAT applications, where ZBDDs play an important role as a slave tool interacting with a master SAT solver. ZBDDs are used as a low-level core technique to compress the internal clause database of the SAT solver.

To this respect, one of the main problems of SAT solvers is the exponential growth of the underlying CNF data structure. As introduced in Section 2, ZBDDs are particularly suited to represent sparse sets, and, for this peculiarity, they have been adopted to compact the CNF database. Each CNF clause is encoded as a single path (from root to the 1-terminal) in a ZBDD representing the set of all clauses. The number of ZBDD paths thus equals the number of clauses in the clause database with a possibly relevant memory compression rate. Implementing CNF manipulations as symbolic operations may exploit size reduction in terms of time complexity, too.

More specifically, a ZBDD encoding uses two variables for every original CNF variable, one for each literal. A clause is encoded by a characteristic function, where the generic variable $x_i = 1$ if the corresponding literal belongs to the clause, otherwise $x_i = 0$.

Example 2.4 As an example the set of clauses [4]:

$$\begin{aligned} S_{cl} \;=\; & (a \vee b \vee \overline{c}) \wedge (a \vee b \vee c) \wedge (a \vee c \vee d) \wedge (\overline{a} \vee b \vee \overline{c}) \wedge \\ & (\overline{a} \vee c \vee \overline{d}) \wedge (\overline{a} \vee c \vee d) \wedge (\overline{b} \vee \overline{c} \vee \overline{d}) \wedge (\overline{b} \vee \overline{c} \vee d) \end{aligned}$$

is represented as a characteristic function $S_{cl}(a, \overline{a}, b, \overline{b}, c, \overline{c}, d, \overline{d})$, depending on 8 variables.

Unfortunately, symbolic encoding and manipulation of CNF clauses imply major modifications in the SAT procedures themselves, often preventing some key performance enhancements of standard explicit SAT.

4.1 ZBDDs for Symbolic Davis-Putnam Resolution.

ZBDDs were first adopted by Chatalica *et al.* [13] to encode CNF clauses and perform a symbolic version of the DP procedure. They argued that the implicit representation is able to overcome the space complexity of DP. Since the amount of clauses generated at each resolution step (variable quantification) may grow exponentially, DP is in fact acknowledged as generally less memory efficient than DLL. By resorting to ZBDDs, the authors are able to symbolically perform several resolutions in a single step, which allowed them to solve problems (e.g., the Pigeon-Hole and Urquhart) previously known as exponential for SAT.

4.2 ZBDDs for Symbolic DLL. Aloul *et al.* [4, 5] use ZBDDs for a similar technique, applied to DLL search. The methodology works as follows.

The clause database is firstly implicitly represented as a ZBDD. Then the original steps of the SAT algorithms are implemented on this structure. "Decision", for example, is implemented by adding the ZBDD for the one-literal clause to the original ZBDD. Analogously, implicit operators on ZBDDs implement unit clause Boolean Constant Propagation and Backtrack search on sets of clauses.

The procedure shows an exponential advantage over the corresponding explicit version on selected test cases.

4.3 ZBDDs for Breadth-First SAT. Finally, Motter *et al.* [32] use ZBDDs to implement a SAT decision process based on Breadth-First Search (*BFS*). As in previously mentioned approaches, ZBDDs allow overcoming space limitations, which are in this case connected with explicit state enumerations in queues and priority queues, typical of BFS procedures.

4.4 Conclusions. All the above methods show how ZBDDs can deliver relevant memory reduction. Moreover, memory compression provides proportional gains on time performance, thanks to symbolic manipulations. As a general remark, ZBDDs are supposed to provide higher speed-ups on classes of problems showing high compression of CNF clauses.

Albeit the evident gains experienced on the proposed benchmarks, the generality of ZBDD based approaches is questionable, as most of the recent improvements attained on explicit SAT solvers, such as conflict diagnosis and recursive learning, are not implemented with ZBDDs, and it is even not clear how to possibly exploit them symbolically. As a consequence, any serious alternative approach in this direction should, at least, address this issue.

5. Decision Diagram Preprocessing and Circuit-Based SAT

We now present approaches requiring a looser integration between BDD and SAT tools. Symbolic BDD based manipulations can be seen as a sort of front-end normalization or pre-processing, in order to reduce the amount of work required by further SAT reasoning. This has been proposed in the domain of combinational as well as sequential verification (Bounded and Unbounded Model Checking). We first describe two works where preprocessing with a BDD like structure is employed in order to normalize and simplify circuit representations before their translation to CNF formulas. Then, we introduce two approaches, doing both a BDD-like circuit preprocessing and a special-purpose SAT algorithm operating directly over the circuit structure. We finally describe preliminary BDD manipulations, as a way to produce redundant information to be exploited by the SAT tool for better search space pruning.

5.1 BED Preprocessing

Abdulla *et al.* [1] and Williams *et al.* [38] propose two similar approaches to solve unbounded Model Checking.

Both works adopt a preliminary Decision Diagram representation for the circuit. BEDs are adopted in [38], whereas a functionally similar structure (Reduced Boolean Circuit) is used in [1]. After that, both approaches do intermediate symbolic manipulations. They eventually convert such a representation to a proper format to perform the final proof. In [1] the authors use a SAT solver to perform the final proof, while in [38] the authors either use a SAT solver or convert their representation into BDDs.

The two works use similar variable quantification rules within image and set inclusion/intersection operations on BEDs. Since BEDs are an intermediate representation between circuit and canonical BDDs, symbolic manipulations done in this phase can also be viewed as local BDD-like function normalizations and simplifications over a combinational unrolling of the circuit, and circuit representations of state sets.

Using the BED representation, the algorithms can take advantage of their built-in reduction rules and the sharing of isomorphic sub-formulas.

In the experimental result section, both papers present data on peculiar circuits, namely a multiplier and a barrel shifter. Figure 2.12 reports a diagrammatic representation of the experimental results obtained by [38] for the multiplier case. Circuit c6288, a 16 bit x 16 bit combinational multiplier, is compared with a shift-and-add sequential implementation. The verification consists in checking that corresponding outputs coincide when the shift-and-add version has finished its computation. The graph plots the CPU time of the different methods (the proposed method using BED, and the tools NuSMV, SMV, and FixIt) as a function of the number of outputs verified

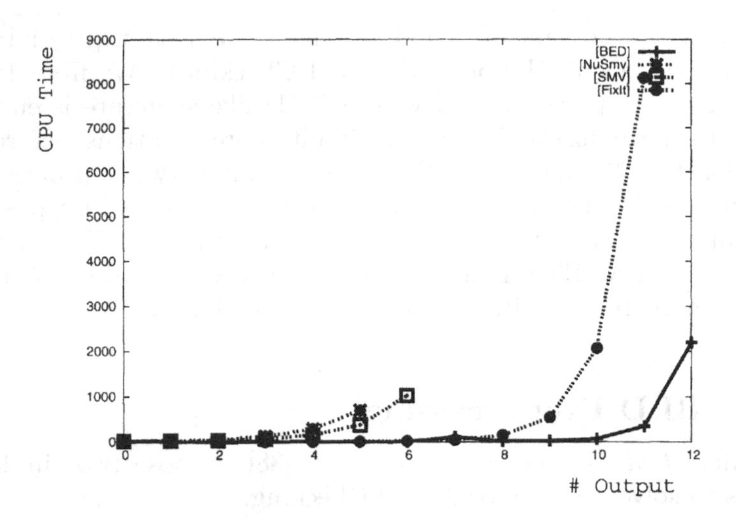

Figure 2.12. Multiplier results.

5.2 Circuit-Based SAT

It is generally accepted that CNF representation is a source of major benefits for SAT solvers, as this simple, general and regular data structure is the base for important optimizations within the SAT engine. In contrast, the CNF format has some drawbacks:

- Any structural information of the circuit, often of crucial importance, is completely lost.
- In many problems, a large number of instances of SAT has to be solved for each circuit. Hence, mapping a given problem description to CNF can represent a significant percentage of the overall running time.

With the purpose of addressing the above problems, SAT can be done directly on the circuit structure (after preliminary BDD-like manipulations) instead of working on CNF representations.

5.2.1 BDD Sweeping and SAT. Kuehlmann *et al.* [25] present an algorithm for Boolean reasoning based on BDDs, structure transformations, and SAT procedure. The approach includes an initial BDD pre-processing, followed by a SAT procedure working directly on the circuit structure.

BDD pre-processing is basically oriented to remove structural redundancies. It works on a BDD-like canonical AND/INVERTER graph (*AIG*) representation of the problem. AIGs use two-input AND vertices, and

INVERTER attributes on the edges. Similar to BDDs, they support efficient structural reductions as a hash-table is used to remove structural redundancy during construction. Moreover, a two-level lookup scheme allows converting any local four-input sub-structure into a canonical representation, effectively removing local redundancy. As redundancies are usually very common in practical problems because of different sources (e.g., the language parsing and processing, the miter structure, the repeated invocations of Boolean reasoning on similar problems, etc.) the reduction is very efficient.

The structural SAT strategy implements a standard search procedure directly on the AND/INVERTER graph representation of the circuit. It attempts to find a set of consistent assignments that satisfy (or violate) the property. The underlying circuit structure enables several optimizations. For instance, propagating implications on circuit topology, which is further enhanced by canonical AIG representation (using only one type of vertex function allows an efficient propagation of logic implication). BDD sweeping [26] also incrementally simplifies the graph structure, due to its ability to efficiently find vertices that are functionally identical or complemented. If a pair of equivalent vertices is found, the algorithm merges them and rebuilds their fan-out structure forward.

In the overall algorithm BDD sweeping and SAT solvers are applied in an intertwined manner as reported in Figure 2.13.

```
BDDSWEEPINGSATSEARCH (f, BDDLimit, SATLimit)
  if (f = 0)
    return (Satisfiable)
  if (f = 1)
    return (Unsatisfiable)
  while (HEAP ≠ ∅)
    result ← BDDSWEEP (HEAP, BDDLimit)
    if (result ≠ Undecided)
      return (result)
    result ← SATSEARCH (f, SATLimit)
    if (result ≠ Undecided)
      return (result)
    BDDLimit ← BDDLimit + DeltaBDDLimit
    SATLimit ← SATLimit + DeltaSATLimit
```

Figure 2.13. BDD Sweeping and SAT Search.

f represents the problem to solve. BDD sweeping is applied first. Basically, it incrementally compresses the structure from the inputs toward the outputs. This effectively reduces the search space for the subsequent SAT search, which is run on f. If none of the two approaches solves the problem, memory and time limits are increased and the process iterates.

Termination comes when the problem is solved or there is no more room for reductions. The algorithm also keeps a heap with "hidden" BDDs, i.e., all BDDs exceeding the current size limit. This reduce the number of BDD re-computations of already computed BDDs, when BDD sweeping is re-invoked with a higher size limit.

As far as experimental results are concerned, the authors compare their algorithm against the original BDD sweeping [26] technique. Results are impressive, showing benefits in terms of memory and time sometimes of two order of magnitude. Unfortunately, no direct comparison with a generic SAT solver is provided. It is thus uneasy to quantify the benefits coming from circuit SAT over generic SAT.

5.2.2 SAT on BEDs. An approach following a similar track has been proposed by Williams *et al.* [37], who operate a full verification procedure on BEDs. Given a BED for a formula, one way of proving its satisfiability is to convert the BED to the corresponding BDD. In this case canonicity would obviously imply size explosion. As a consequence, the proposed approach is to use an upper level SAT-like partitioning, so that BED sizes (and more importantly the BDD sizes after BED to BDD conversion) are kept under control.

The authors first build a BED representation of the satisfiability problem. Then they use a combination of SAT-based variable decision and BED to BDD conversion.

If the BED is large enough (which is always the case for non trivial problems), the procedure (a top level recursive split variable selection procedure called *BedSat*) performs a case splitting, and problem partitioning, similar to CNF based SAT decisions. The algorithm works on BEDs, so variable selection heuristics are changed, and no BCP is done, but a similar process exploiting direct implications through BEDs rewriting rules. Using BEDs the selection is obtained by pulling a variable to the root.

Whenever a BED is small enough after a set of variable decisions, the algorithm builds a BDD (starting from the BED) and checks whether the result is different from 0.

Experiments show that the method is faster than both pure BDD construction and the straight use of satisfiability solvers such as GRASP and SATO. A diagrammatic representation of these results, for the ISCAS'85 circuits, is reported in Figure 2.14. Plots are not represented when the computation could not be completed within the resource limits. Data, albeit interesting, are not conclusive as they mainly involve "small" benchmarks also often dealt with more "traditional" methods.

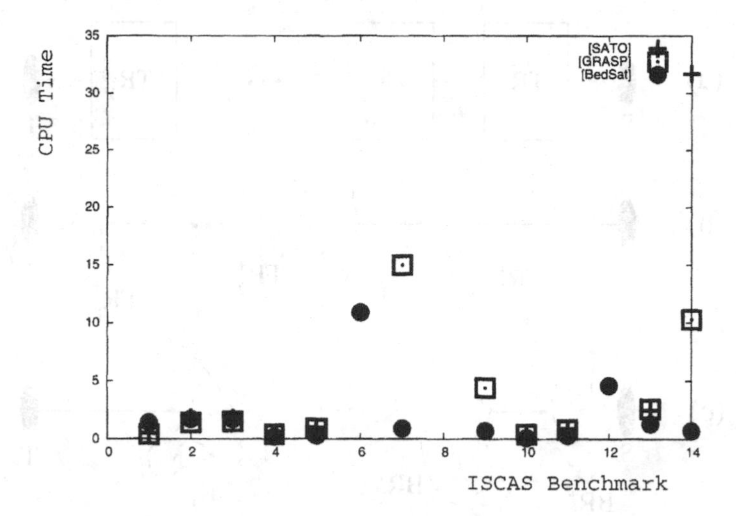

Figure 2.14. BDD and SAT comparison.

5.3 Preprocessing by Approximate Reachability

Cabodi *et al.* [12] explore an alternative way to make BDD-based and SAT-based tools cooperate. Their target is to improve the efficiency of SAT-based BMC with the help of affordable BDD-based pre-processing.

It is well known (see for example [14, 20]) that "approximate" BDD traversals may deal with much larger circuits than "exact" ones, at the expense of exactness. Moreover, as the degree of approximation can be trimmed, it is always possible to trade-off memory and time with the accuracy of the result. Unfortunately nothing comes for free and the limit of approximate techniques in verification is that they are not complete, i.e., over-approximate reachability can prove correctness, but it cannot disprove it.

The driving idea of [12] is to complement the initial over-approximate BDD traversal with a final SAT-solver search. In other words, BDDs are used as a redundant extra information, in order to prune and focus the search. The high-level flow is represented in Figure 2.15.

First of all, as introduced in Section 2, a combinational unrolling of the circuit representation (TR) of length k is generated (see Figure 2.15 (a)). By adding the expressions for S and T, and performing a proper variable re-labeling for TR a propositional formula is stored in CNF format. After that, a BDD manipulation phase computes the characteristic functions of the over-approximate state sets. This can be done in the forward or backward directions (see Figure 2.15 (b) and (c)) or mixing them both. Then the estimate is combined, as an additional constraint, with the

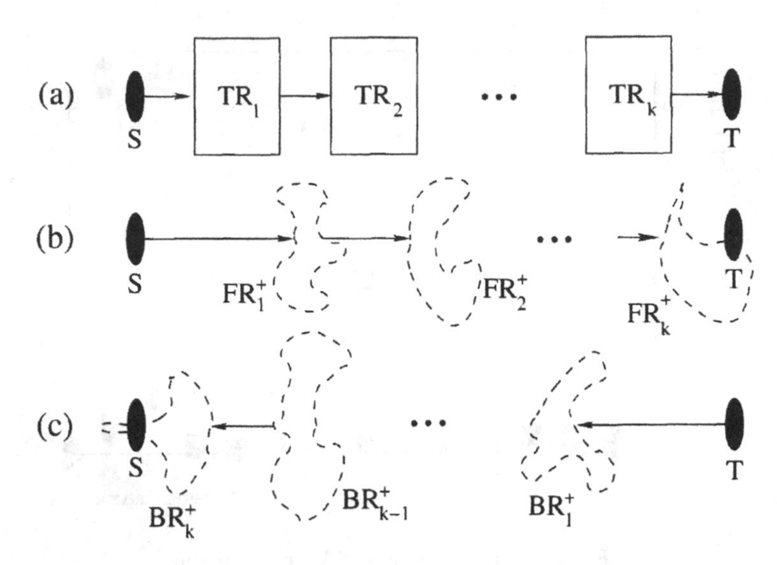

Figure 2.15. (a) Standard combinational unrolling for SAT-based BMC; (b) Approximate forward traversal from S to T; (c) Approximate backward traversal from T to S.

previously generated combinational unrolling. Finally, the SAT-engine is run on the resulting problem to solve it.

The main target in this case is to obtain an efficient pruning on the SAT solver search space. The method shows best effects with increasing bounds (sequential depth), and it has the advantage of requiring a loose integration with the SAT solver.

6. Using SAT in Symbolic Reachability Analysis

This last section presents approaches using SAT tools within inner steps of symbolic image and/or pre-image computations. All the presented approaches share the idea of representing state sets by means of their characteristic function, which is a typical paradigm of BDD based verification.

We first present a BDD based verification technique using SAT procedures to drive BDD partitioning and manipulations within image computations.

Then we describe two approaches performing full image/pre-image computations within a SAT tool: The former approach does not use BDDs at all, but we include it in our presentation for two reasons:

- It proposes CNF clauses and SAT solving procedures to fully substitute BDDs in one of their most popular applications.

- It is a reference work for the latter approach, where a BDD-like decomposition is adopted for state set compressed representation.

6.0.1 BDDs at SAT leaves. Gupta *et al.* [22, 23] perform BDD based reachability analysis by using a SAT procedure within symbolic image computation. They call their approach *BDDs at SAT Leaves.*

More specifically, they use BDDs to represent state sets and a CNF formula to represent the transition relation. Symbolic image of a state set is computed by exhaustive SAT search of all solutions within the space of primary input, present and next state variables. However, rather than using SAT to enumerate each solution all the way down to a leaf, image switches to BDD-based computations at certain intermediate points within the SAT decision tree. This is done as a trade off between space complexity of BDDs and time complexity of full SAT enumeration.

In a sense, this approach can be regarded as SAT providing a disjunctive decomposition for image computation into many sub-problems, each of which is handled symbolically using BDDs.

The image of a From state set (expressed by a BDD) is computed by a top level SAT procedure working on the CNF formula expressing TR, with the following major modifications:

- Decisions not satisfying the From domain set are forbidden. This restriction mechanism (called *BDD Bounding*) is achieved by taking the conjunction of the From set and the transition relation.
- Whenever a satisfying assignment is found, present state variables are not recorded (they are implicitly quantified out) so that solutions are represented within the next state space.
- The SAT procedure stops when the whole space has been visited, not after finding any single solution (as standard SAT procedures).

One naive approach of performing *BDD Bounding* would be enumerating each complete SAT solution up to the leaf of the search tree, and then check if the solution satisfies the given BDD. Since this would obviously be quite inefficient, the authors interleave decisions of present and next state variables, and they dynamically check present state assignments against the BDD of From (with no added cost).

A key aspect of the approach is choosing when to switch from SAT decision tree to BDD symbolic image. In fact, the main claim of the authors is that their approach drastically relaxes the BDD blow-up problem, since the only BDDs required are for state set representation, not in monolithic, but in partitioned form. The upper SAT decision tree works as a disjunctive "chronological" partitioning of an image problem is subproblems.

An attractive aspect of this solution is the ability to drive partitioned BDD manipulations with SAT-based control. However, a few inherent limitations are represented by BDD partition overlapping, uneasy manipulation of partitions with different variable ordering, tuning thresholds and parameters that control the SAT to BDD switch. Moreover, the reported experiments mainly target a direct comparison with standard BDD-based symbolic reachability analysis. It is thus again uneasy to quantify the benefits coming from dovetailing SAT and BDDs over pure SAT methods.

6.0.2 SAT-Based Symbolic Image and Pre-image. McMillan [29] introduced a fully novel approach to perform unbounded CTL model checking completely within a SAT tool. The approach is based on a CNF quantifier elimination procedure. While the top level algorithm is basically the same as used in BDD-based CTL model checking, sets of states are represented as CNF formulas rather than with BDDs. This required a modification of the SAT solver in order to be able to perform the key operation of quantifier elimination, i.e., the enumeration of all solutions. Whenever a satisfying assignment is found, instead of terminating, the algorithm seeks for new solutions, until full exploration of the solution space.

A key factor for efficiency is search space pruning, based on learning. Whenever a new solution is found (and the corresponding clause added to the solution set), the algorithm augments the clause database with the so called *Blocking Clause*, which is in conflict with (i.e., it rules out) the satisfying assignment.

Experimental results show that this technique can compete with BDD based model checkers and in some cases outperform them. The author compares BDDs and SAT on publicly available model checking problems derived from the compositional verification of one unit of a commercial microprocessor design. All the checked formulas are invariants and they are all true, i.e., no counter-example is generated. Figure 2.16 reports these results by plotting the performance of the SAT-based method against the performance of the BDD-based method. The author observes that while the total run time is much smaller for the BDD-based techniques, for most individual problems, the SAT-based method is faster (in some cases by two order of magnitude). To conclude, this is a pretty good results for SAT-based methods as it is likely to be much room for improvements in the SAT-based methods than in the BDD-based ones. At least these results seems to suggest it would be a good policy to devote at least a short time to SAT-based methods before trying BDD-based approaches.

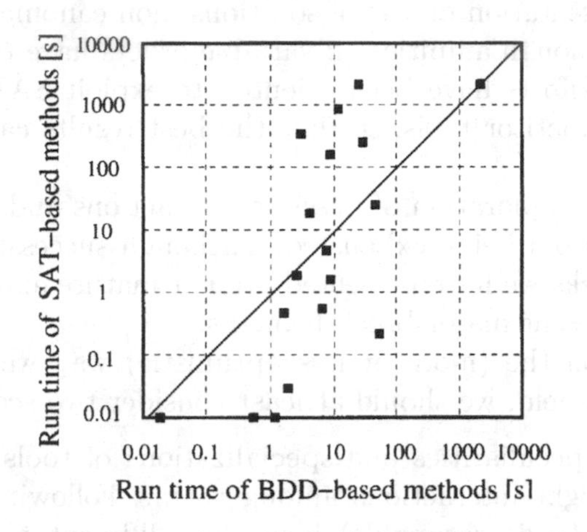

Figure 2.16. Microprocessor Verification Results.

A related approach is presented in [35]. Here pre-image is achieved through an ATPG based SAT search. Success-driven learning is performed as in [29], but BDD like graphs are used to store state sets. BDDs are not canonical ROBDDs and they are not used for symbolic manipulations. They rather resemble Free-BDDs [6] (BDDs with different orderings along different paths), mainly exploited to obtain (possibly exponential) size compaction over CNF representation of state sets.

The presented algorithm prunes redundant search space due to overlapped solutions and construct a FBDD on the fly. At the end, the FBDD becomes the representation of the pre-image set.

The experimental results focus on pre-image computation for large ISCAS benchmarks.

7. Conclusions, Remarks and Future Works

In this chapter we have overviewed most of the works mixing SAT and BDD techniques in order to make them cooperate. After presenting the main differences and affinities between them, we have followed a classification scheme based on different application frameworks and interaction schemes adopted for SAT-BDD interaction.

Let us now briefly draw some conclusions and lessons learned from the experiences we have described.

First of all, we should definitely agree that SAT algorithms and symbolic BDD manipulations are indeed different, and they are located at opposite points in the space of solution strategies. Differences emerge on most relevant choices characterizing the approaches: Time vs. space complexity, depth-first vs. breadth-first visits, explicit decision tree vs. symbolic representation of sets of solutions, non canonicity vs. canonicity, etc. Even though affinities on variable orders have been considered, most research efforts have been oriented to exploit SAT and BDDs as alternative competitor tools, so that the best results can be reaped by both of them.

However, the exploration of possible interactions and intertwined operations is far from being exhausted. Although successful stories exist, most of the works we have presented are not mature and/or not general enough to represent major breakthroughs.

Depending on the (more or less optimistic) way we look at future research in this field, we should at least consider two scenarios.

- Different peculiarities and specializations of tools dominate, and prevent tight interactions among them. Following the generally accepted (and reasonable) idea that different tools have different classes of tractable problems, research and engineering efforts will be directed at better characterizing and predicting tractable problem instances and dynamically switching between available engines.

- Tools are different and highly specialized, but they are activated on partitions and/or sub-tasks of large problems, so that their degree of interaction is tighter.

For obvious reasons, let us concentrate on the latter possibility, where using BDDs under given (tractable) size thresholds, then switching to SAT decision procedures, is definitely an attractive solution. This is the leading idea of several presented approaches, seeking for an optimal trade-off between time and memory limitations.

Dynamic self-tuning and scalability are key features for widespread applicability and robustness. To this extent, Kuehlmann *et al.* [25] show advanced aspects such as tight integration of technologies and iterative increase of size and backtrack limits.

Other interesting, still not fully explored trends, include:

- Representation and/or manipulation of state sets by SAT solvers. McMillan [29] opens an alternative field, state sets instead of combinational unrolling, for SAT manipulations, whereas [12] shows that state sets are an important additional information to constrain SAT search space.

- Improved BDD manipulations can be achieved by resorting to search techniques using SAT. In addition to explicit search, which can leverage size explosion (as in [23]), sophisticated backtracking schemes, learning and implication handling are mode specialized and powerful than standard depth-first visits of BDD graphs.

- Relaxing canonicity is known as a way to face BDD size explosion. CNF clauses are an extreme (non canonical) choice, while intermediate solutions, like for instance BEDs [24], still lack a solid and fully developed technology (especially in sequential verification).
- Low level symbolic manipulations are an attractive issue for SAT search. ZBDDs [4, 5, 13, 32] thus far just opened a track. A crucial question is how symbolic manipulations, albeit their potential compactness, may keep pace with the dramatic performance improvements strongly related to the simple and regular structure of CNF clauses.

References

[1] P. A. Abdulla, P. Bjesse, and N. Eén. Symbolic Reachability Analysis based on SAT-Solvers. In *TACAS 2000 - Tools and Algorithms for the Construction and Analysis of Systems*, 2000.

[2] F. A. Aloul, I. L. Markov, and K. A. Sakallah. Faster SAT and Smaller BDDs via Common Function Structure. In *Proc. Int'l Conf. on Computer-Aided Design*, San Jose, California, November 2001.

[3] F. A. Aloul, I. L. Markov, and K. A. Sakallah. MINCE: A Static Global Variable–Ordering for SAT and BDD. In *Proc. Int'l Workshop on Logic Synthesis*, Lake Tahoe, California, May 2001.

[4] F. A. Aloul, M. N. Mneimneh, and K. A. Sakallah. Backtrack Search Using ZBDDs. In *Proc. Int'l Workshop on Logic Synthesis*, Lake Tahoe, California, May 2001.

[5] F. A. Aloul, M. N. Mneimneh, and K. A. Sakallah. ZBDD–Based Backtrack Search SAT Solver. In *Proc. Int'l Workshop on Logic Synthesis*, Lake Tahoe, California, May 2002.

[6] J. Bern, C. Meinel, and A. Slobodová. Some Heuristics for Generating Tree–like FBDD Types. *IEEE Transactions on CAD*, 15(1):127–131, January 1996.

[7] A. Biere, A. Cimatti, E. M. Clarke, M. Fujita, and Y. Zhu. Symbolic Model Checking using SAT procedures instead of BDDs. In *Proc. 36th Design Automat. Conf.*, pages 317–320, New Orleans, Louisiana, June 1999.

[8] P. Bjesse, T. Leonard, and A. Mokkedem. Finding Bugs in an Alpha Microprocessor Using Satisfiability Solvers. In Gérard Berry,

Hubert Comon, and Alan Finkel, editors, *Proc. Computer Aided Verification*, volume 2102 of *LNCS*, pages 454–464, Paris, France, July 2001. Springer-Verlag.

[9] R. E. Bryant. Graph–Based Algorithms for Boolean Function Manipulation. *IEEE Transactions on Computers*, C–35(8):677–691, August 1986.

[10] J. R. Burch, E. M. Clarke, D. E. Long, K. L. McMillan, and D. L. Dill. Symbolic Model Checking for Sequential Circuit Verification. *IEEE Transactions on CAD*, 13(4):401–424, April 1994.

[11] G. Cabodi, P. Camurati, and S. Quer. Can BDDs compete with SAT solvers on Bounded Model Checking? In *Proc. 39th Design Automat. Conf.*, New Orleans, Louisiana, June 2002.

[12] G. Cabodi, S. Nocco, and S. Quer. Improving SAT-based Bounded Model Checking by Means of BDD-based Approximate Traversals. In *Proc. Design Automation & Test in Europe Conf.*, pages 898–903, Munich, Germany, March 2003.

[13] P. Chatalic and L. Simon. ZRes: the old DP meets ZBDDs. In *Proc. 17th Conf. of Autom. Deduction (CADE)*, 2000.

[14] H. Cho, G. D. Hatchel, E. Macii, B. Plessier, and F. Somenzi. Algorithms for Approximate FSM Traversal Based on State Space Decomposition. *IEEE Transactions on CAD*, 15(12):1465–1478, December 1996.

[15] A. Cimatti, E. M. Clarke, F. Giunchiglia, and M. Roveri. NuSMV: a new Symbolic Model Verifyer. In *Proc. Computer Aided Verification*, volume 1633 of *LNCS*, pages 495–499. Springer-Verlag, July 1999.

[16] F. Copty, L. Fix, R. Fraer, E. Giunchiglia, G. Kamhi, A. Tacchella, and M. Y. Vardi. Benefits of Bounded Model Checking at an Industrial Setting. In Gérard Berry, Hubert Comon, and Alan Finkel, editors, *Proc. Computer Aided Verification*, volume 2102 of *LNCS*, pages 435–453, Paris, France, July 2001. Springer-Verlag.

[17] M. Davis, G. Logemann, and D. Loveland. A Machine Procedure for Theorem-Proving. *Journal of the ACM*, 5:394–397, 1962.

[18] M. Davis and H. Putnam. A Computing Procedure for Quantification Theory. *Journal of the ACM*, 7:201–215, 1960.

[19] E. Goldberg and Y. Novikov. BerkMin: a Fast and Robust SAT-Solver. In *Proc. Design Automation & Test in Europe Conf.*, pages 142–149, Paris, France, February 2002.

[20] S. G. Govindaraju, D. L. Dill, A. Hu, and M. A. Horowitz. Approximate Reachability Analysis with BDDs using Overlapping Projec-

tions. In *Proc. 35th Design Automat. Conf.*, pages 451–456, San Francisco, California, June 1998.

[21] J. F. Groote and F. Zantema. Resolution and binary decision diagrams cannot simulate each others polynomially. Technical report, Utrecht University, 2000.

[22] A. Gupta, Z. Yang, P. Ashar, and A. Gupta. SAT–Based Image Computation with Application in Reachability Analysis. In *Proc. Formal Methods in Computer-Aided Design*, volume 1954 of *LNCS*, Austin, TX, USA, 2000.

[23] A. Gupta, Z. Yang, P. Ashar, L. Zhang, and S. Malik. Partition–Based Decision Heuristic for Image Computation using SAT and BDDs. In *Proc. Int'l Conf. on Computer-Aided Design*, San Jose, California, November 2001.

[24] H. Hulgaard, P. F. Williams, and H. R. Andersen. Equivalence checking of combinational circuits using boolean expression diagrams. *IEEE Transactions on CAD*, July 1999.

[25] A. Kuehlmann, M. K. Ganai, and V. Paruthi. Circuit-based Boolean Reasoning. In *Proc. Design Automat. Conf.*, Las Vegas, Nevada, June 2001.

[26] A. Kuehlmann and F. Krohm. Equivalence Checking Using Cuts and Heaps. In *Proc. 34th Design Automat. Conf.*, pages 263–268, Anaheim, California, June 1997.

[27] J. P. Marques-Silva and K. A. Sakallah. GRASP – A New Search Algorithm for Satisfiability. In *Int'l Conference on Tool with Artificial Intellingence*, 1996.

[28] K. McMillan. *Symbolic Model Checking.* Kluwer Academic, Boston, Massechusset, 1994.

[29] K. L. McMillan. Applying SAT Methods in Unbounded Symbolic Model Checking. In Ed Brinksma and Kim Guldstrand Larsen, editors, *Proc. Computer Aided Verification*, volume 2404 of *LNCS*, pages 250–264, Cophenagen, Denmark, 2002.

[30] S. I. Minato. Zero–Suppressed BDDs for Set Manipulation in Combinational Problems. In *Proc. 30th Design Automat. Conf.*, pages 272–277, Dallas, Texas, June 1993.

[31] M. Moskewicz, C. Madigan, Y. Zhao, L. Zhang, and S. Malik. Chaff: Engineering an Efficient SAT Solver. In *Proc. 38th Design Automat. Conf.*, Las Vegas, Nevada, June 2001.

[32] D. B. Motter and I. L. Markov. A Compressed Breadth-First Search for Satisfiability. 2002.

[33] M. Prasad, P. Chong, and K. Keutzer. Why is ATPG easy? In *Proc. 36th Design Automat. Conf.*, pages 22–28, New Orleans, Louisiana, June 1999.

[34] S. Reda, R. Drechsler, and A. Orailoglu. On the Relation between SAT and BDDs for equivalence checking. *Int'l Symposium on Quality of Electronic Design (ISQED)*, pages 394–399, 2002.

[35] S. Sheng and M. Hsiao. Efficient Preimage Computation Using A Novel Success-Driven ATPG. In *Proc. Design Automation & Test in Europe Conf.*, pages 822–827, Munich, Germany, March 2003.

[36] T. E. Uribe and M. E. Stickel. Ordered binary decision diagrams and the Davis-Putnam procedure. In *ICCCL*, volume 845 of *LNCS*, pages 34–49. Springer-Verlag, 1994.

[37] P. F. Williams, H. R. Andersem, and H. Hulgaard. Satisfiability Checking Using Boolean Expression Diagrams. In *TACAS 2001 - Tools and Algorithms for the Construction and Analysis of Systems*, IT University of Copenhagen, April 2001.

[38] P. F. Williams, A. Biere, E. M. Clarke, and A. Gupta. Combining Decision Diagrams and SAT Procedures for Efficient Symbolic Model Checking. In E. Allen Emerson and A. Prasad Sistla, editors, *Proc. Computer Aided Verification*, volume 2102 of *LNCS*, pages 124–138, Chicago, Illinois, July 2000. Springer-Verlag.

[39] L. Zhang and S. Malik. The Quest for Efficient Boolean Satisfiability Solvers. In Ed Brinksma and Kim Guldstrand Larsen, editors, *Proc. Computer Aided Verification*, volume 2404 of *LNCS*, pages 17–36, Cophenagen, Denmark, 2002.

Chapter 3

EQUIVALENCE CHECKING OF ARITHMETIC CIRCUITS

Dominik Stoffel
Dept. of Electrical and Computer Engineering, University of Kaiserslautern, Germany
stoffel@eit.uni-kl.de

Evgeny Karibaev
Dept. of Electrical and Computer Engineering, University of Kaiserslautern, Germany
karibaev@eit.uni-kl.de

Irina Kufareva
Department of Radio Physiks, Tomsk State University, Tomsk, Russia
irene@molsoft.com

Wolfgang Kunz
Dept. of Electrical and Computer Engineering, University of Kaiserslautern, Germany
kunz@eit.uni-kl.de

Abstract Although equivalence checking technology has matured greatly during the last few years and designs with millions of gates can be handled, some specific problems remain to be difficult. Formal verification of arithmetic circuits, especially if multiplication is involved, is one of these problems. In this chapter we analyze origin and nature of this problem. We then review the most important research contributions targetting equivalence checking of arithmetics. Specifically, we outline techniques exploiting arithmetic functional properties and techniqes based on binary decision diagrams, both on the bit and word levels. We also report on our own experiments with *Multiplicative Binary Moment Diagrams* (*BMDs). Finally, we introduce a new pragmatic approach to equivalence checking of arithmetic circuits. It extracts from a gate-level netlist an *arithmetic bit-level* representation of the circuit. Verification is carried out on this representation using a simple arithmetic calculus rather than a Boolean one. We show experimental results for successfully equivalence checking a large number of industrial multipliers as well as other circuits implementing more complex arithmetic expressions.

Keywords: Equivalence checking, arithmetic verification, datapath verification, multiplier verification, decision diagrams, arithmetic bit level

R. Drechsler (ed.), Advanced Formal Verification, 77-123.

1. Introduction

Modern circuit design flows increasingly employ formal verification techniques in order to ensure quality and reduce time-to-market by avoiding bug-related design iterations. Equivalence checking has become a regular step in the flow. Modern tools for equivalence checking are capable of verifying designs with millions of gates in very short times.

The great success of equivalence checking technology is due to numerous research advancements in this field during the past decade. An important idea on which a typical equivalence checker is based is to exploit *structural similarity* between the two circuit models being compared [1, 2]. Structurally similar circuits contain a lot internal nodes implementing equivalent circuit functions. These *internal equivalences*, sometimes called *cut points* [3] can be used to efficiently break the verification problem down into smaller ones as has been explored by several researchers [4, 2, 5, 3, 1, 6].

In many applications of equivalence checking, the two circuit models being compared exhibit a great amount of structural similarity. For example, during logic synthesis, the transformations of a gate netlist into another gate netlist are of fairly local scope so that many equivalent internal functions remain. However, in some cases, structural similarity is not given. An important example that occurs often in practice is the verification of arithmetic circuits. The problem occurs when an RT-level (*register transfer level*) specification of a circuit must be compared against a gate-level implementation, e.g., when verifying the correctness of a logic synthesis step. Figure 3.1 illustrates this case. The verification engines in a typical equivalence checker all operate on gate-level circuit models. Hence, in order to compare an RTL specification with a gate-level implementation, the frontend of the verification tool first has to generate a gate-level representation of the specification. The process is similar to the logic synthesis step that produced the implementation. The two gate netlists can then be compared by the backend engines to verify equivalence or produce a counterexample. When the design contains arithmetic functions, this approach is bound to fail. The problem is that the two gate netlists hardly contain any structural similarity at all. The reason for this lies in the great flexibility when implementing arithmetic functions.

In general, arithmetic functions in digital circuits, such as addition, subtraction, multiplication and division, are implemented using addition as the base function. Subtracting a number X in two's complement notation from a number Y, for example, is implemented by inverting all bits of X, adding 1, and adding Y. Multiplication is also based on

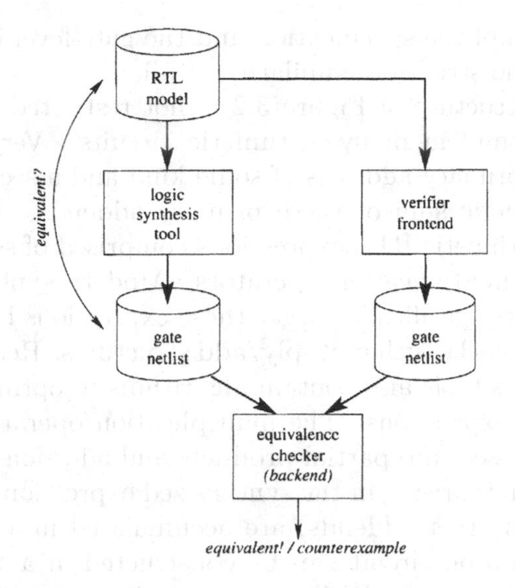

Figure 3.1. Equivalence checking of an RTL model against a gate netlist

addition. Hardware multipliers are most often composed of two stages (Fig. 3.2). In the first stage, the partial products are generated from two operand vectors, X and Y. The way the partial products are generated depends on whether signed or unsigned numbers are processed, and whether or not Booth recoding is used. The partial products are inputs to the second stage which is an addition circuit. In the sequel we will call the inputs to an addition circuit *primary addends*. The addition circuit adds the primary addends up to produce the final result $Z = X \cdot Y$. The implementation of this addition circuit can be chosen from a variety of architectures differing in performance or area requirements. Most common implementations are an array of *carry-save adders (CSA)* or a *Wallace tree*. The great variety of possible implementations

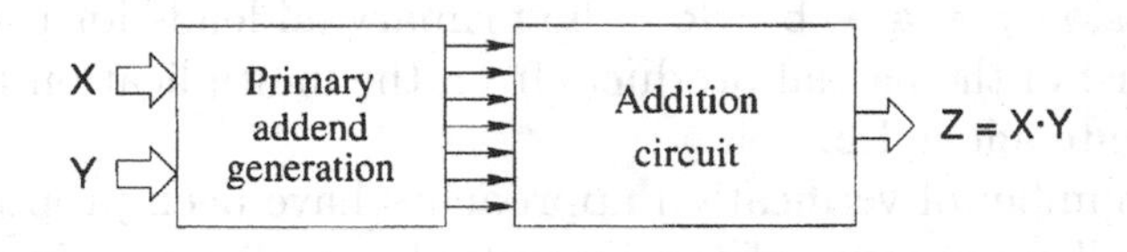

Figure 3.2. Basic multiplier structure

for both stages of the circuit makes it impossible for the verifier front-end to constructively reproduce the output of the synthesis tool. Hence, the gate-level version of the specification and the gate-level implementation exhibit little or no structural similarity at all.

The general structure of Figure 3.2 is not restricted merely to multipliers but is found in many arithmetic circuits. Very often, a first stage generates primary addends of some kind and a second stage computes the arithmetic sum of these primary addends. Also, many designs contain arithmetic RTL expressions comprised of several addition, subtraction and multiplication operators. Modern synthesis tools and module generators specifically target these expressions by offering more general building blocks such multiply/add structures. Recently, commercial logic synthesis tools also contain algorithms to optimize specifically these arithmetic expressions. The multiplication operators of the RTL code are decomposed into partial products and addition circuits. Then, all primary addends arising in the synthesized expression (including partial products and other addends) are accumulated in a single addition circuit. This addition circuit can be constructed in a "globally" optimal structure, e.g., using a Wallace tree architecture. Figure 3.3 shows

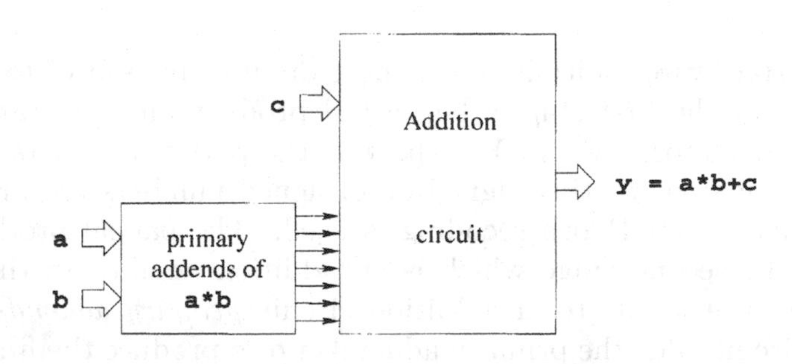

Figure 3.3. More general arithmetic circuit example

an example of such a more general circuit structure generated from the RTL expression `y = a * b + c`. The primary addends for the addition circuit consist of the partial products from the multiplication `a * b` and of the separate addend `c`.

A great number of verification approaches have been proposed in the past. We will cover some of the important contributions in this chapter. All approaches attempt to exploit specific knowledge about the nature of the verification problem of checking the equivalence of arithmetic circuits. In Section 2 we discuss approaches that are based on special properties of the arithmetic *functions* implemented by a circuit. Section 3 discusses approaches that apply Binary Decision Diagrams exploiting *structural* information about the designs. In Section 4 we give an introduction to word-level decision diagrams which are an attempt to bridge the gap between the Boolean domain of gate-level circuits and the word-level domain of RTL models. Finally, Section 5 introduces an approach that exploits knowledge about the typical internal structure of arithmetic circuits as shown in Figures 3.2 and 3.3. It extracts from the gate-level netlist of a circuit an arithmetic bit-level representation which serves as the basis for verification algorithms.

2. Verification Using Functional Properties

An approach for the verification of arithmetic functions which is based on a standard equivalence checking engine was proposed by Fujita [7]. Some arithmetic functions such as multiplication, square and cube functions have special properties which can be expressed as *recurrence equations*. For a multiplier, a recurrence equation is

$$f(x+1, y) = f(x, y) + y \tag{3.1}$$

The only function that satisfies this equation (and an additional boundary condition $f(0, y) = 0$) for all x and all y is the multiplication function $f(x, y) = x \cdot y$. Instead of verifying the equivalence of a specification and an implementation of multiplication, we verify whether the implementation satisfies the recurrence equation. The corresponding setup is shown in Figure 3.4. In this setup, a miter is constructed of the circuits in

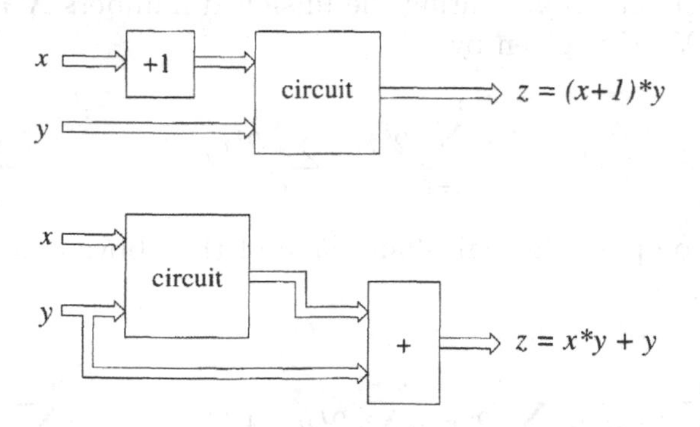

Figure 3.4. Verification of a multiplier using a recurrence equation

the upper part and the lower part of Fig. 3.4. The upper part realizes the left-hand side of Equation 3.1. It consists of an incrementor which adds 1 to the operand x, and of the circuit presumed to be a multiplier. The lower part implements the right-hand side of the equation and is composed of the circuit under verification and of an adder. The circuit being checked is a multiplier if and only if the outputs z of the the two circuits are equivalent.

The idea now is to use a standard equivalence checking engine based on structural similarities. However, from the general structure of Figure 3.4, it is unlikely to find many equivalent signals in the two circuits to be compared. This can be somewhat improved by performing successive case splits on the bits of operand x. For example, if the least significant bit x_0 is set to 0, the incrementer degenerates to an inverter for x_0 and to identity functions for the remaining bits x_i of the operand. In this case, many internal cut points can be found between the two designs and an equivalence check succeeds. The next case split is performed by setting $x_0 = 1$ and $x_1 = 0$. The incrementor degenerates to two inverters and identity functions for the remaining bits. This leads to fewer equivalent functions and to a harder verification subproblem. Nevertheless, the verification problem is broken down. In their experiments, the authors were able to verify the multiplier C6288 from the ISCAS-85 benchmark set in less than 12 minutes on a Sparc20.

The major drawback of this interesting approach is that for the circuit to be checked, a recurrence equation must exist and it must be known. This hampers automation of the verification task.

A related approach has been pursued in [8] and [9] for verifying multipliers. It uses relationships of the following kind. Consider a multiplier for unsigned numbers. Let the operands $\underline{x}$ and $\underline{y}$ be two n-bit and m-bit vectors, respectively, representing the unsigned numbers X and Y. The product $Z = X \cdot Y$ is given by

$$Z = \sum_{i=0}^{n-1} 2^i x_i \cdot \sum_{j=0}^{m-1} 2^j y_j \qquad (3.2)$$

We can decompose the right-hand side of the above equation in the following way:

$$\underbrace{\sum_{i=0}^{n-1} 2^i x_i \cdot \sum_{j=0}^{m-1} 2^j y_j}_{n \times m \text{ multiplier}} = \underbrace{\sum_{i=0}^{n-2} 2^i x_i \cdot \sum_{j=0}^{m-1} 2^j y_j}_{(n-1) \times m \text{ multiplier}} + 2^{n-1} x_{n-1} \cdot \sum_{j=0}^{m-1} 2^j y_j \qquad (3.3)$$

This relationship is used to verify the implementation of the multiplier. Note that the first term of the right hand side of Equation 3.3 corresponds to the multiplication of an $(n-1)$-bit number with a m-bit number. If we have a circuit implementing an n-bit $\times$ n-bit multiplier, we can easily obtain an $(n-1)$-bit $\times$ n-bit multiplier by setting the most significant bit (MSB) x_{n-1} of operand $\underline{x}$ to 0. The remaining terms represent a partial product and an adder. The idea of [8, 9] is to construct a miter circuit representing this equation as shown in Figure 3.5. If we

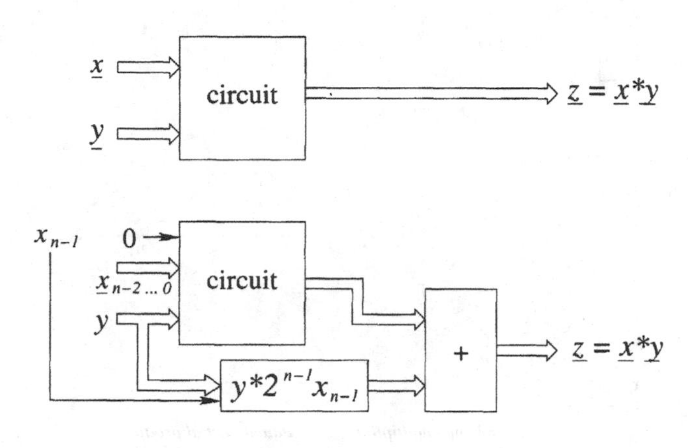

Figure 3.5. Inductive equivalence checking of multipliers

prove the equivalence of the outputs of this miter, then we have shown that the circuit is an $n \times m$ multiplier provided setting $x_{n-1} = 0$ makes it an $(n-1) \times m$ multiplier. In other words, the construction checks whether the multiplier includes a correct addition of the partial product formed with x_{n-1}. The correctness of the $(n-1) \times m$ multiplier is again checked by setting up the same construction, this time x_{n-2} being the most significant bit which is set to 0. The complete verification consists of n steps in each of which one more operand bit x_i is set to 0 and the equivalence check using the miter of Figure 3.5 is performed.

Each miter is verified using a standard cutpoint-based equivalence checker. From the structure of Figure 3.5 one should think that a large number of internal equivalences exist since the lower circuit is derived from the upper circuit by only setting an input to 0 and appending some logic primarily at the outputs. However, structural similarities exist only in special cases and only if additional information is used from the structure of the circuit under verification [8].

Figure 3.6 shows a simplified example of a miter setup for the induction-based approach. The upper half of the figure displays part of a multiplier adding four partial products for the addition tree of output column z_3. (For simplicity, we completely disregard adding of carry signals from previous columns). In the lower part of the figure the x_3-

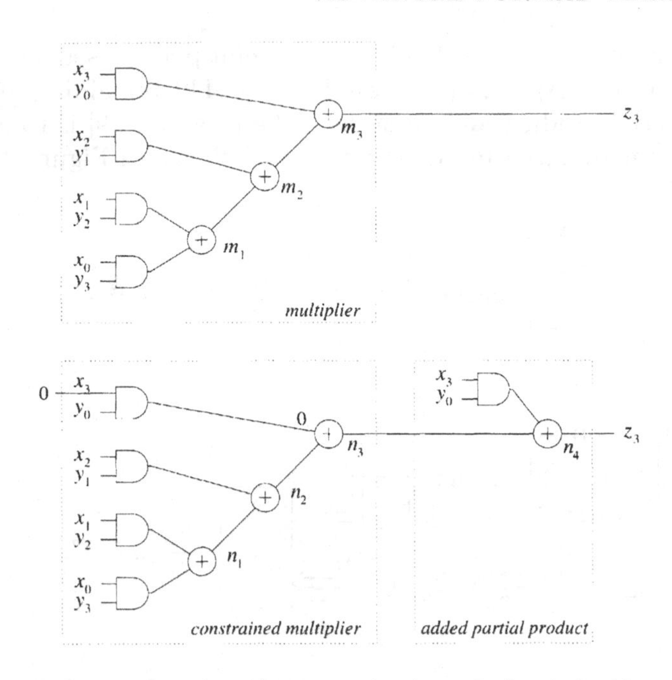

Figure 3.6. Induction-based verification – simple equivalence checking problem

input of the multiplier is set to 0. The partial product x_3y_0 is added to the output of the multiplier. Both circuits are checked to be equivalent using an equivalence checker. The shown problem is easy for the checker since many equivalences exist, e.g., $m_1 \equiv n_1$, $n_2 \equiv n_3 \equiv m_2$, and $m_3 \equiv n_4 \equiv z_3$. The situation changes if a different architecture is used for the addition circuit. Consider Figure 3.7. In this case, the partial products are added in a different order. No internal equivalences exist.

In the case of array architectures, it is possible to select the operand bits x_i to be set to 0 in the reverse order of their accumulation in the multiplier so that mostly easy situations as in Figure 3.6 occur. By choosing the operand bits in this order, the successive iterations of adding partial products "reconstruct" the multiplier step by step. However, for Wallace tree [10] architectures, it is not possible to find such a "good" order of the operand bits since it is not possible to reconstruct the tree structure by simply appending partial product additions. In these cases, the authors of [8, 9] resort to combining their approach with the verification technique of [11]. However, for Wallace-tree architectures, the induction-based verification approach still has problems with robustness [9].

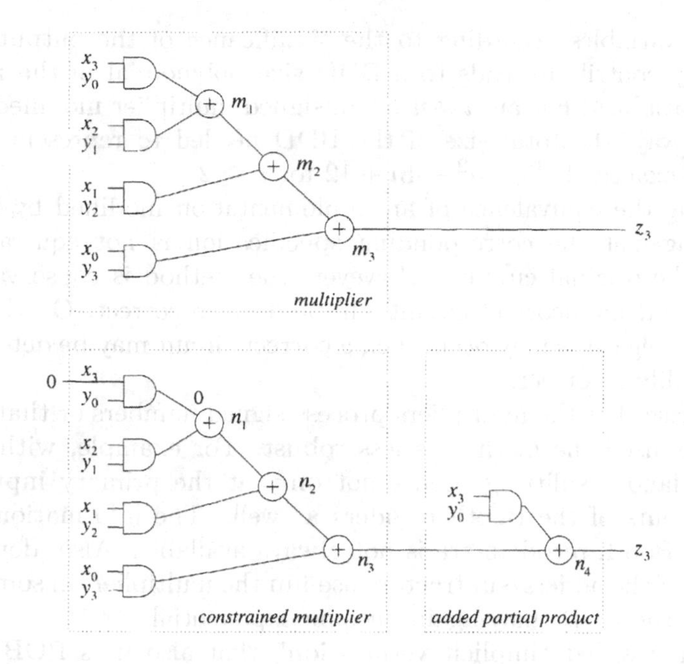

Figure 3.7. Induction-based verification – hard equivalence checking problem

3. Bit-Level Decision Diagrams

Already in Bryant's famous paper of 1986 [12] introducing Reduced Ordered Binary Decision Diagrams (ROBDDs) it was proven that multiplier circuits cannot be represented efficiently by BDDs. Regardless of the variable order, the ROBDD size is always exponential in the number of variables. Hence, the straightforward approach of constructing "monolithic" BDDs for the circuit outputs and comparing them for isomorphism does not work even for small multipliers. Nevertheless, ROBDDs and their derivatives have been applied to the verification of arithmetic circuits by using smarter approaches.

One of the first techniques was an equivalence checking approach reported by Burch [13]. The method is described in detail for unsigned combinational multipliers that compute products by adding together partial products bits $x_i \wedge y_j$ for all inputs x_i and y_j. The basic idea is to apply fanout splitting to the primary inputs of the multiplier. Each fanout branch of an operand bit is replaced by a new input variable. In order to verify the modified circuit, a corresponding specification circuit has to be generated to which fanout splitting has also been applied. The resulting circuits are no longer multiplier circuits. Hence, a variable ordering resulting in a moderate BDD growth may exist. In fact, ordering the input variables according to the significance of the output bit to which they contribute leads to a BDD size polynomial in the number of input variables. For an $n \times n$ bit unsigned multiplier modified in the described way, the total size of the BDD needed to represent all the outputs is exactly $4n^3 - 6n^2 - 4n + 12$ for $n \geq 2$.

Checking the equivalence of an implementation modified by fanout-splitting against the corresponding specification is not equivalent to checking the original circuits. However, the method is conservative in the sense that no incorrect circuit will be deemed correct. On the other hand, false negatives may occur, i.e., a correct circuit may be determined to be possibly incorrect.

In the case that the multipliers process signed numbers or that Booth recoding is used the method is less robust. For example, with Booth recoding, fanout splitting occurs not only at the primary inputs but at the outputs of the Booth recoders as well. The information about where the Booth recoders are is not always available. Also, depending on details of the adders/subtractors used in the multiplier, in some cases the size of the BDDs may again become exponential.

A method called "implicit verification" that also uses ROBDDs to verify multipliers and other arithmetic circuits was proposed by Stanion [11]. The method exploits what the author called "structural dependence". The circuit is partitioned into sub-circuits by backwards traversal from the circuit outputs, starting with the least significant output bit. As an example, Figure 3.8 shows the partitioning of a multiplier circuit producing an n-bit output word into n subcircuits P_0, P_1, $\ldots$, P_{n-1}. Each subcircuit P_k has as inputs a number of primary inputs corresponding to the operand bits x_i, y_j, and a set of signals K_{k-1} originating in the previous sub-circuit (fanout sets). If the size of the

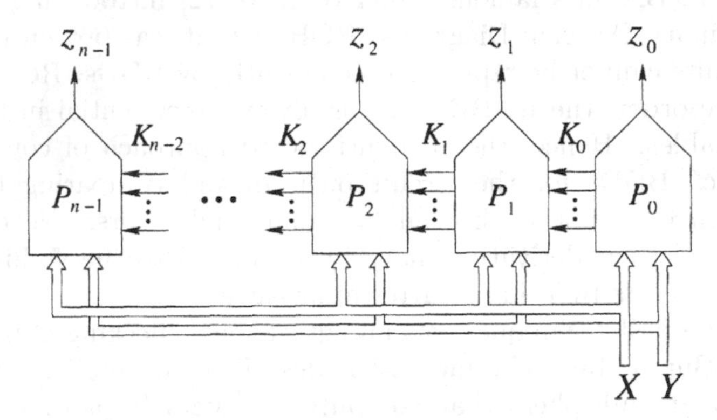

Figure 3.8. Partitioning a circuit for implicit verification

fanout sets is relatively large compared to the number of inputs to the subcircuits, the circuit is said to exhibit structural dependence that can be exploited in verification.

Structural dependence is exploited in the following way. The circuit is verified output bit by output bit, i.e., each primary output z_i of the implementation is compared with the corresponding output z'_i of the specification. The primary pairs of output bits are proven in ascending order, i.e., the least significant bit z_0 is proven first and the most significant bit z_{n-1} is proven last. (In the following, a symbol without a prime (′) refers to the specification, a symbol with prime refers to the implementation.)

For each pair of output bits, an ROBDD is constructed computing the XOR of the functions implemented by the corresponding subcircuits P_i and P'_i. This ROBDD represents the characteristic function $g = z_i \oplus z'_i$ of all counterexamples for the equivalence of the circuit outputs. These counterexamples, however, contain assignments to the variables in the fanout sets K_{i-1} and K'_{i-1} which are not satisfiable if the circuits are equivalent. In other words, the requirement $z_i \neq z'_i$ places constraints on the input variables and on the variables of the fanout set K_{i-1}. In order to prove that z_i and z'_i are equivalent, these constraints have to be proven to be contradictory. This is done by additionally taking the results of previously proven output bits into account. When proving output bits z_i and z'_i, the equivalence of the previously targetted output bits z_{i-1} and z'_{i-1} has already been established. We can form the characteristic function $h_{i-1} = \overline{z_{i-1} \oplus z'_{i-1}}$ representing all assignments to the inputs of the subcircuits P_{i-1} and P'_{i-1} for which the output bits are equivalent. Also, the complete input/output behaviour of these verified subcircuits can be represented by characteristic functions χ_{i-1} and ${\chi'}_{i-1}$, respectively. These functions evaluate to 1 for all assignments to the inputs and outputs of the subcircuit which are valid with respect to the corresponding subcircuit's function. Specifically, the characteristic functions evaluate to 0 for all assignments to the primary inputs and to the variables in the fanout sets K_{i-1} and K'_{i-1} that violate the subcircuits' behaviour or the equivalence of the circuit outputs. By forming a conjunction of function g with the characteristic functions h_{i-1}, χ_{i-1} and ${\chi'}_{i-1}$ all counterexamples violating subcircuits P_{i-1} and P'_{i-1} are excluded. This may immediately lead to an evaluation of the conjoined functions to 0 which proves the equivalence of the output bits z_i and z'_i. If, however, the conjoined functions do not evaluate to 0, additional constraints have to be added resulting from the next lower output bits z_{i-2} and z'_{i-2}. This can be continued for all previously proven circuit outputs. Table 3.1 shows the pseudo-code for the described approach (taken from [11] and slightly adapted to the notations in this chapter).

```
function imp_ver() {
    partition circuits;
    for i = 0 to n − 1 do {
      g = z_i ⊕ z_i';
      for j = i − 1 downto 0 do {
        h_j = NOT(z_j ⊕ z'_j);
        g = g · χ_j · χ_j' · h_j;
        if g = 0 then break;
      }
      if g ≠ 0 return INEQUIVALENT;
    }
  return EQUIVALENT;
}
```

Table 3.1. Implicit verification algorithm for structurally dependent circuits

Using implicit verification, the author of [11] was able to check the equivalence of structurally dissimilar circuits such as a Wallace tree multiplier against a CSA array multiplier. Empirically, the memory requirements for representing the ROBDDs seem to double with every 4 bits of operand size. Also the CPU times increased accordingly. The author reports verification of 32×32 multipliers in about six hours of CPU time (as of 1999) requiring 130 MBytes of memory. These results are much better than what could be achieved by constructing monolithic BDDs for the circuit functions. However, the seemingly exponential dependence of the problem complexity on the number of input variables makes this technique less robust for larger operand bit widths.

4. Word-Level Decision Diagrams

Struggling with the problems encountered in arithmetic circuit verification, researchers looked for alternatives to bit-level graph representations. The invention of word-level decision diagrams are an attempt to exploit specific knowledge about the circuits other than structural information. At the same time, it is important to maintain the advantages of compact representation and easy manipulation, since these properties have made decision diagrams in the (bit-level) Boolean domain very successful. The idea is to make use of the fact that arithmetic circuits implement arithmetic functions which should be dealt with in the (integer) arithmetic domain rather than the (Boolean) logic domain. Datapath circuits process bundles of binary signals that encode words of data. This abstraction from binary-valued Boolean signal values to word-level values such as integers or real numbers is the basis of word-level decision diagrams. Many approaches have been pursued [14, 15, 16, 17, 18, 19, 20] in this direction. The basic ingredient of all these approaches is the use of *pseudo-Boolean* functions which are functions over Boolean variables but having non-Boolean ranges, such as integers or real numbers. In this section we review the basic concepts of word-level decision diagrams. From a practical point of view, among all approaches proposed so far, so-called *multiplicative binary moment diagrams (*BMDs)* [18] have proven to be the most popular. We will look at these decision diagrams and algorithms for their synthesis in more detail.

4.1 Pseudo-Boolean functions and decompositions

A *pseudo-Boolean function* is a function $f : B^n \rightarrow Z^m$. It maps vectors of binary values to vectors of integer numbers. Figure 3.9 shows an example of such a function (taken from [18]).

x_1	x_2	$f(x_1, x_2)$
0	0	8
0	1	-12
1	0	10
1	1	-6

Figure 3.9. Example [18] of a pseudo-Boolean function $f : B^2 \rightarrow Z$

The concept of a pseudo-Boolean function has several benefits. From an application point of view, it allows to link the Boolean domain of binary-valued signals processed by gate-level logic circuits with the arithmetic word-level domain of more abstract circuit models. From the viewpoint of designing data structures and algorithms for function representation and manipulation, it offers to use function decompositions similar to the ones on which binary decision diagrams are based such as the Shannon decomposition. This is key for graph representations of such functions.

The basic idea is to consider B as a subset of Z. An input variable x evaluates to either 0 or 1. Therefore, Boolean negation (INV) of a variable x can be expressed by integer arithmetic as $(1 - x)$ and the

Boolean conjunction operation (AND) of l variables $x_1, x_2, \ldots, x_l$ can be expressed by the arithmetic product $x_1 \cdot x_2 \cdot \ldots \cdot x_l$. Such an (integer arithmetic) product term evaluates to 1 if and only if all variables evaluate to 1, i.e., they are equivalent to the corresponding Boolean "product" of the Boolean variables.

Using these integer arithmetic transforms for the Boolean operations, we can effectively decompose pseudo-Boolean functions pointwise, just as in the Boolean domain. For example, the function of Figure 3.9 can be expressed in a "minterm" normal form by multiplying each minterm with the corresponding function value and adding up all terms:

$$\begin{aligned} f(x_1, x_2) = \; & 8 \cdot (1-x_1) \cdot (1-x_2) \\ & + (-12) \cdot (1-x_1) \cdot x_2 \\ & + 10 \cdot x_1 \cdot (1-x_2) \\ & + (-6) \cdot x_1 \cdot x_2 \end{aligned} \tag{3.4}$$

Simplifying this equation yields a more compact algebraic representation of the function:

$$f(x_1, x_2) = 8 + 2x_1 - 20x_2 + 4x_1x_2 \tag{3.5}$$

In the Boolean domain, the Shannon expansion is used for function decompositions and representation as *binary decision diagrams* (BDDs) [12]. A Boolean function f is expanded in terms of a variable x by $f = \overline{x} \cdot f_{\overline{x}} + x \cdot f_x$, where $\cdot$ and $+$ denote Boolean conjunction and disjunction, respectively. The terms $f_{\overline{x}}$ and f_x are the functions resulting when variable x is set to 0 and 1, respectively. They are called the negative and positive *cofactors* of f with respect to variable x. The Shannon expansion can be generalized to pseudo-Boolean functions by replacing Boolean operators with integer arithmetic operators. If $\cdot$, $+$ and $-$ denote the arithmetic operators of multiplication, addition and subtraction, the Shannon decomposition for pseudo-Boolean functions is:

$$f = (1-x) \cdot f_{\overline{x}} + x \cdot f_x \tag{3.6}$$

In the same way, other decompositions such as the *Reed-Muller* or Davio decomposition of the Boolean domain can be generalized for pseudo-Boolean functions. For representing Boolean functions, these decompositions have been used for so-called *functional decision diagrams* (FDDs) [21, 22]. Equations 3.7 and 3.8 are the positive and negative Davio decompositions generalized to pseudo-Boolean functions, respectively.

$$f = f_{\overline{x}} + x \cdot (f_x - f_{\overline{x}}) \tag{3.7}$$

$$f = f_x - (1 - x) \cdot (f_x - f_{\overline{x}}) \tag{3.8}$$

Equations 3.7 and 3.8 can be easily verified to be equivalent with the Shannon expansion by the reader.

Depending on the type of decomposition used, different types of word-level decision diagrams can be developed. The Shannon decomposition of pseudo-Boolean functions (Eq. 3.6) leads to so-called *Multi-Terminal Binary Decision Diagrams* or MTBDDs [16], also called Algebraic Decision Diagrams (ADDss) in [17]. Figure 3.10 shows an example of an MTBDD for the function of Figure 3.9.

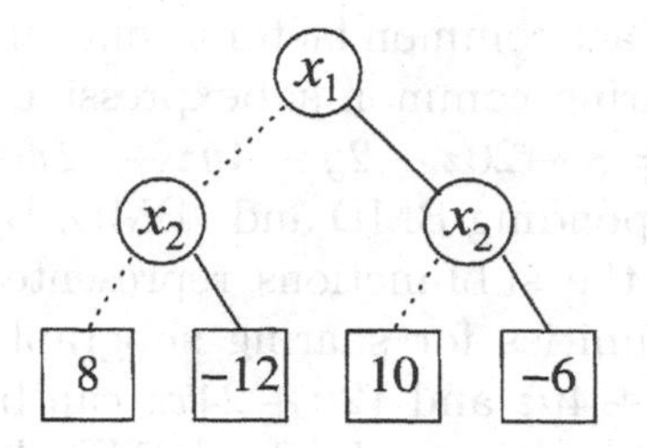

Figure 3.10. Example of MTBDD [18]

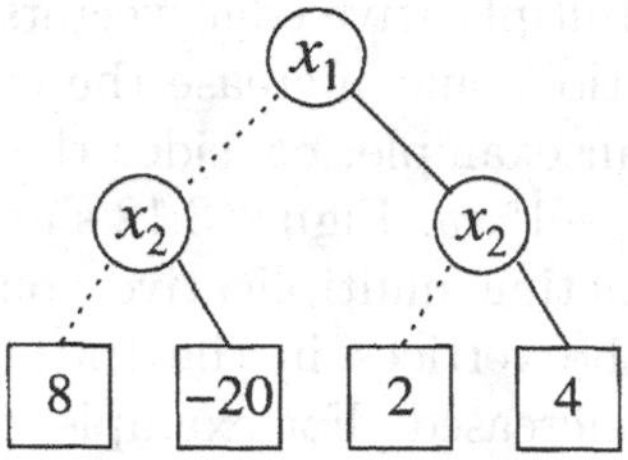

Figure 3.11. Example of BMD [18]

Using the positive Davio decomposition of Eq. 3.7, we obtain a different type of decision diagram called *Binary Moment Diagram* (BMD). The name comes from the fact that we can view the pseudo-Boolean function f of Eq. 3.7 as a linear function in x. The term $(f_x - f_{\overline{x}})$ is equivalent to the partial derivative of f with respect to x. It is called the *linear moment* [18] of f. Analogously, the term $f_{\overline{x}}$ represents the part of f that is independent of x. It is therefore called the *constant moment* of f.

Applying the positive Davio decomposition recursively on the function of Figure 3.9 yields the BMD shown in Figure 3.11. This can be verified by rearranging and factoring the terms of Eq. 3.5:

$$f(x_1, x_2) = (8 - 20x_2) + x_1(2 + 4x_2) \tag{3.9}$$

Note that MTBDDs and BMDs are graph representations of pseudo-Boolean functions that are not likely to be compact in the general case. The integer values of the range of the represented function are stored in the terminal values. Sharing of isomorphic sub-graphs is less frequent than for bit-level decision diagrams because of the multiplicity of possible terminal values. To allow more compact representations, extensions have been proposed based on *edge weights*. Each edge in the decision diagram is marked by an integer weight. An *additive weight* of an edge is interpreted by adding the weight to the function to which this edge is incident. Introducing additive edge weights to MTBDDs lead to so-called *Edge-Valued Binary Decision Diagrams* (EVBDDs) [14]. *Multiplicative edge weights* are multiplied with the function to which an edge is incident. Extending BMDs with multiplicative edge weights yields so-called *BMDs. For representing arithmetic functions, *BMDs are suited much better than EVBDDs. Therefore, the following sections discuss *BMDs and their synthesis in more detail.

4.2 *BMDs

Multiplicative edge weights allow to extract common factors from subfunctions and increase the chance for sharing common subexpressions. As an example, consider the function $y = 8 - 20z + 2y + 4yz + 12x + 24xz + 15xy$. Figure 3.12 shows the corresponding BMD and *BMD. By extracting multiplicative constants from the subfunctions represented by the vertices in the BMD, the opportunities for sharing subgraphs are increased. For example, the terms $2y + 4yz$ and $12x + 24xz$ can be factored as $2y(1 + 2z)$ and $12x(1 + 2z)$, respectively. In the *BMD the subgraph representing $1 + 2z$ is shared, after constant factors have been extracted and placed as weights to the edges in the graph. (Weights are represented by square boxes located next to the edges).

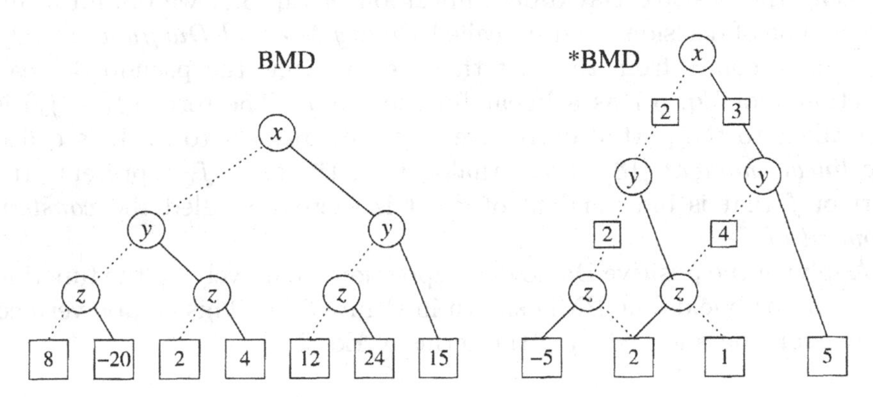

Figure 3.12. Example of BMD versus *BMD [18]

By obeying special rules when extracting and manipulating edge weights, *BMDs can be maintained canonical. One rule, for example, is that the edge weights for two branches leaving a vertex be relatively prime. It is also required, that the integer 0 never appears as edge weight but only as a terminal value, and that when a node has a branch to 0, the other branch has weight 1.

*BMDs are very well suited to represent arithmetic functions such as word-level addition, subtraction, multiplication and exponentiation. The *BMDs of these operations are very compact, especially if compared to their bit-level BDD counterparts. It is a remarkable feature of *BMDs to be capable of efficiently representing the multiplication operation which is infeasible to be represented by an ROBDD. Figure 3.13

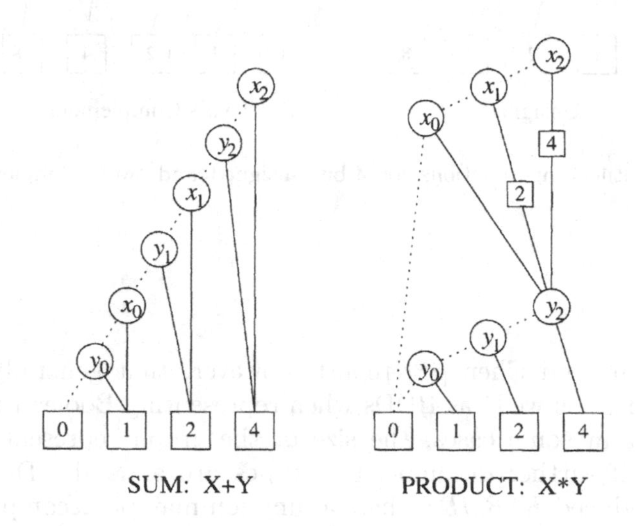

Figure 3.13. Word-level sum and product of 3-bit unsigned numbers [18]

shows the *BMDs for the operations of addition and multiplication, respectively, of two unsigned 3-bit numbers.

Pseudo-Boolean functions and *BMDs represent word-level operations. Digital circuits, however, process individual binary signals using logic gates. The semantics of a set of signals is given by a specific encoding of these signals. When linking bit-level to word-level representations, *encoding* functions are needed. The *BMDs for such encoding functions are also very compact. Figure 3.14 shows the *BMDs for unsigned and two's complement encodings, respectively, of 4-bit data words. A data word is a vector (x_3, x_2, x_1, x_0) of Boolean variables with x_3 being the most signficant bit.

Boolean functions can also be represented by *BMDs, simply by observing that the Boolean values $\{0, 1\}$ are a subset of integers and, hence, can also be represented by a pseudo-Boolean function. However, the representations of Boolean functions by *BMDs and by BDDs are not always equally complex. There are cases of Boolean functions for which *BMDs are much more complex than the BDD representations and vice

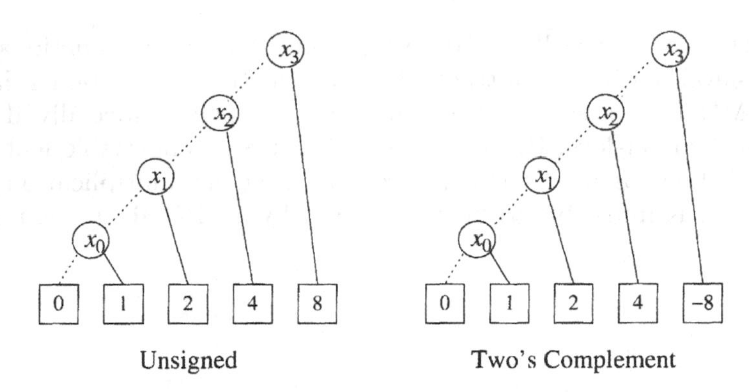

Figure 3.14. Encoding functions for 4-bit unsigned and two's complement numbers [18]

versa. Bryant and Chen [18] report, however, that generally *BMDs behave "almost as well" as BDDs when representing Boolean functions.

Note that in some cases, the size of the graph representation can be reduced if further decomposition types are allowed. Drechsler et al. [20] introduced *K*BMDs* where a function may be decomposed with respect to a variable using either Shannon, positive or negative Davio decomposition (Equations 3.6–3.8). K*BMDs may have both additive and multiplicative edge weights. Chen and Bryant proposed so-called *multiplicative power hybrid decision diagrams (*PHDDs)* [23] which are also based on all three decomposition types. They further employ power-of-two edge weights and complement edges for negation. This allows efficient representations of floating point functions.

4.3 Equivalence Checking Using *BMDs

The task of equivalence checking is to verify whether or not two circuit models called *specification* and *implementation* implement the same function. For datapath circuits, verification using *BMDs seems very attractive. As shown in the previous section, *BMDs are very efficient in representing word-level functions such as number encodings and arithmetic operators. The canonicity property of *BMDs allows us an easy proof of equivalence of two word-level functions by checking the respective *BMDs for isomorphism. However, in most equivalence checking scenarios, at least one of the two circuit models is given as a gate netlist on the bit-level (not on the word-level). Figure 3.15 shows as an example the case where the implementation is given as a netlist consisting of logic gates implementing bit-level Boolean functions, whereas the specification is given on the word-level.

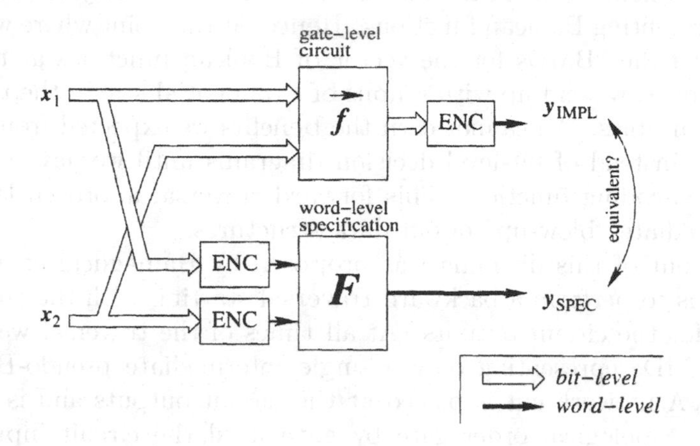

Figure 3.15. Equivalence Checking with *BMDs

In order to verify the equivalence of implementation and specification, it is necessary to link the bit-level domain of logic gates with the word-level domain of *BMDs using encoding functions. Let the circuit in the example of Figure 3.15 process two words of data x_1, x_2 to produce one output word y_{IMPL}. Each data word is a vector of Boolean signals. In order to compare the implementation to the specification, these bit vectors are converted to integers using the encoding functions indicated by the boxes named *ENC*. The equivalence check amounts to checking whether the *BMDs for $ENC(f(x_1, x_2))$ and $F(ENC(x_1), ENC(x_2))$ are isomorphic.

The question is, however, how do we obtain the *BMDs? For the specification side this is easy, since we can assemble more complex arithmetic expressions from the *BMDs of the individual arithmetic operators. For the implementation side, this is more difficult. A straightforward approach is to represent the Boolean functions implemented by the gates in the gate netlist by *BMDs. Moving from gate to gate starting at the primary inputs of the circuit and proceeding in topological order we construct the *BMDs for each internal signal using operations similar to "apply" as known from ROBDDs. When we have reached the primary outputs of the gate-level circuit, we have a vector of Boolean functions represented by *BMDs. Applying the encoding function for the circuit outputs to these *BMDs yields the final *BMD representing the pseudo-Boolean function implemented by the circuit.

This approach works sometimes but fails in other very important cases. Let us assume that function F in Figure 3.15 is the multiplication operation. Remember that *BMDs behave similarly to ROBDDs when representing Boolean functions. Hence, at the point where we have constructed the *BMDs for the vector of Boolean functions at the circuit's outputs, we end up with graphs of exponential size in the number of input variables. We cannot earn the benefits we expected from using word-level instead of bit-level decision diagrams until we have also applied the encoding function. This forward traversal approach leads to an intermediate "blow-up" of our data structures.

A way out of this dilemma was proposed by Hamaguchi et al. [19]. The idea is to perform a backward traversal starting with the encoding function for the circuit outputs. At all times of the traversal we maintain a *BMD representing only a single intermediate pseudo-Boolean function. A network cut is placed at the circuit outputs and is moved in reverse topological order gate by gate until the circuit inputs are reached. The signals on the network cut are the input variables to the intermediate *BMD. Each time the cut frontier is moved across a gate, the variable corresponding to the output of the gate is substituted by the gate function expressed in variables corresponding to the gate inputs. This operation is similar to the "compose" operation known from bit-level decision diagrams.

The goal of this approach is to avoid the intermediate BMD blowup. This blowup can only be avoided if the benefits of the word-level graph representation can be exploited. This seems to be the case in the backward traversal, since at all times only a single *BMD representing a single pseudo-Boolean function is maintained. The authors [19] show very good results for a number of multipliers of different bit widths and architectures.

Although these results were promising, *BMDs are still not regularly applied in industrial verification settings. There are a number of reasons. First, *BMDs are compact if the represented functions are arithmetic. If there is a bug in the circuit, however, the graphs tend to explode so that no counterexample can be determined from them. This is, however, of minor importance. Arithmetic circuits, especially multipliers, are very easy to test using random patterns. If there is a bug in the circuit, a counterexample can very often be determined by simulating just a small number of random patterns.

However, there is a more severe fact hampering the application of *BMDs in industrial practice. Hamaguchi's method [19] does not seem to work well on multipliers generated by commercial synthesis tools. These multipliers are less regular than their textbook counterparts and the automatic determination of a good cut frontier is very difficult. Keim et. al. [24] proved for a certain class of multipliers (unsigned integer Wallace-tree type) that the backward traversal approach is of polynomial space and time complexity. The proof is based on a circuit representation of the multiplier consisting of the AND gates implementing the partial product bits and of full adder cells. Unfortunately, in practice, multiplier netlists are composed of general logic gates and information about the arithmetic sub-components is not available.

We experimented with a large number of multipliers generated by our own tools as well as by commercial generators with the goal to improve *BMD synthesis by backward traversal as proposed by Hamaguchi. However, we were only partially successful. In the following section, we report on these experiments to show where the major problems are.

4.4 Experiments with *BMD synthesis

The general intuition behind preferring a backward traversal over a forward traversal is that the former makes better use of the special characteristics of *BMDs when representing arithmetic functions. Representing a pseudo-Boolean function in the *BMD should always be better than representing a bit-level Boolean function. However, if the network cut is not chosen appropriately, the pseudo-Boolean function implemented by the logic between the cut variables and the (integer) circuit output need not necessarily yield a compact *BMD. The work of [19] suggests a simple topological ordering among the nodes of the circuit when composing the *BMD. In our experiments, this node ordering worked well for text-book multipliers generated by our own tool. However, for industrial multipliers, we were not able to construct *BMDs using simple topological ordering. In this section we report on a number of experiments evaluating where the problems are and how *BMD synthesis could be improved.

It has been pointed out by a number of authors [18, 25, 24] that knowledge about the sub-components of a multiplier is very useful in *BMD construction. However, in a gate netlist this information is not available. In order to successfully build a *BMD, however, information about the circuit structure should be exploited. The goal is to derive as much information about the internal structure of the gate netlist as possible.

As a first improvement over simple topological ordering of nodes to be composed into the *BMD, we experimented with partitioning the gate netlist of a multiplier into non-overlapping cones. Each circuit node belongs to exactly one cone. The cones are identified by marking the transitive fanin of the circuit output bits. The outputs are visited in ascending order of their bit significance in the output word of the multiplier.

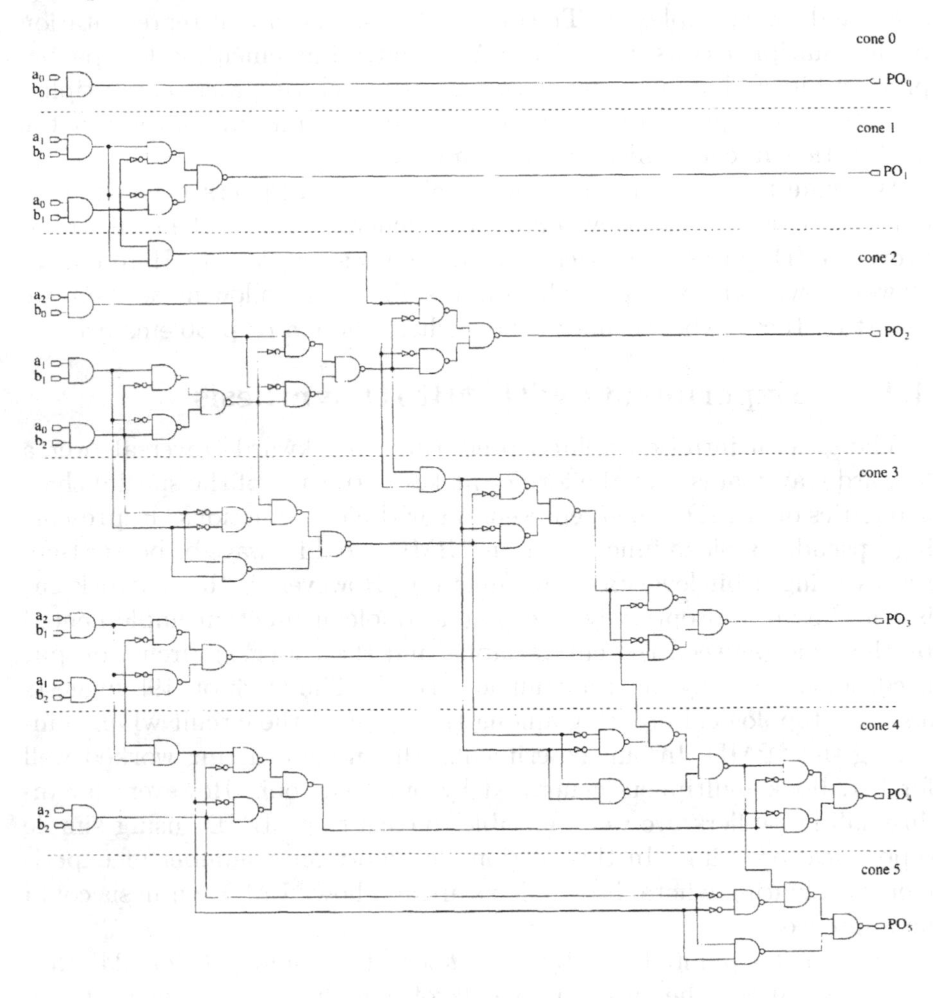

Figure 3.16. Partitioning of a multiplier circuit into non-overlapping cones

Figure 3.16 shows an example for partitioning a multiplier circuit into non-overlapping cones. The partitioning yields circuit regions in the cone of influence of each circuit output containing only gates that are not in the cone of influence of a PO with less significance. As will be explained, construction of the *BMD will be done cone by cone. The cones have two types of inputs: primary inputs and signals from the adjacent cone.

The output of a cone is a primary output bit. The *index* of a cone is the positional index of its output bit in the resulting output bit vector.

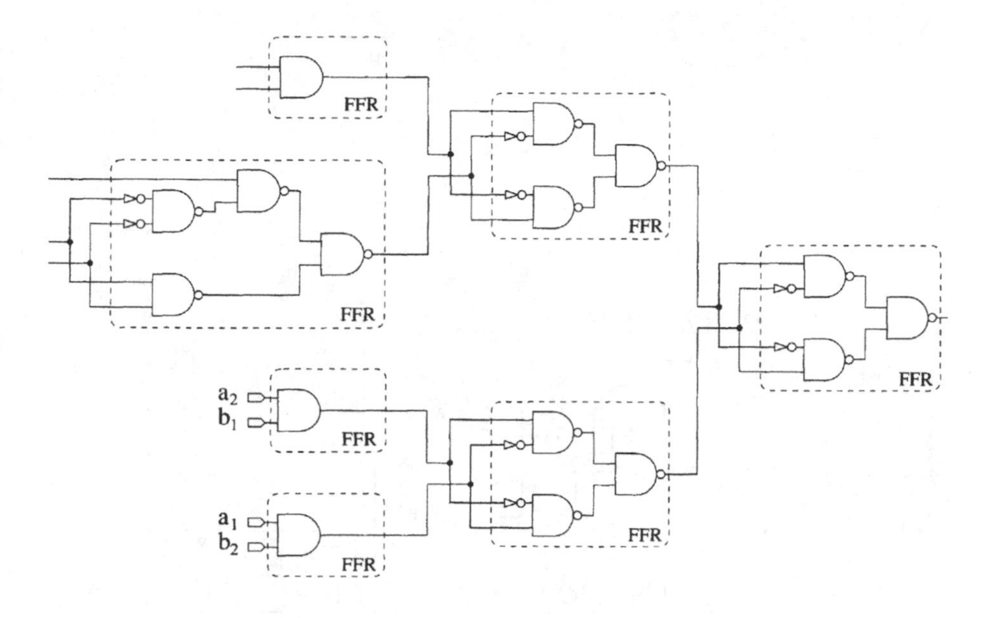

Figure 3.17. Partitioning a cone into fanout-free regions

Cone-wise partitioning has the goal of avoiding intermediate *BMD blow-up. The primary inputs constitute a known and good "cut" frontier. With every cone finished in *BMD synthesis, the cut frontier sucessively contains more and more primary inputs.

Within each cone, we exploit structural information by splitting the circuitry into its fanout-free regions (FFR) [26]. Intuitively, a fanout-free region is a maximal subgraph of the circuit graph such that the subgraph has tree structure. The gates inside an FFR feed exactly one other gate. The inputs to the fanout-free region are either primary inputs or gates with fanout. The output of the fanout-free region is either a primary output or a gate with fanout.

Note that our netlist representation consists only of NAND gates (and inverters), so that also XOR functions may end up in individual fanout-free regions. As an example, Figure 3.17 shows the partitioning of cone 3 of Fig. 3.16 into fanout-free regions.

Each fanout-free region is assigned a *weight*. The weight of an FFR is the number of inputs to the cone in its transitive fanin. The inputs may be either primary inputs or signals from an adjacent cone. The weights

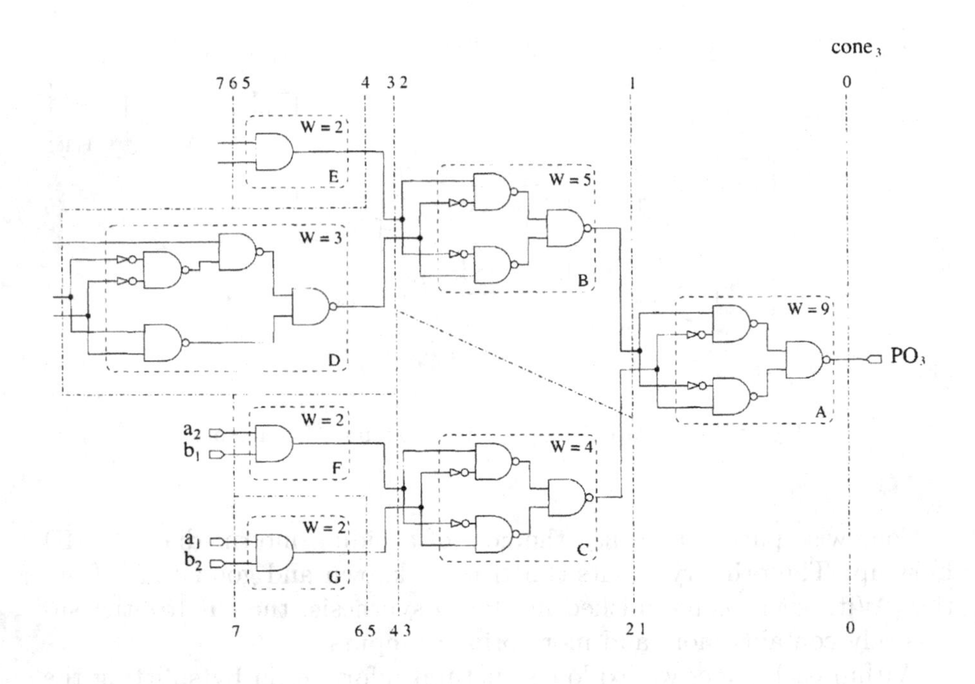

Figure 3.18. Selection of network cuts for backward traversal

are used for selecting the FFRs to be composed in the *BMD synthesis process.

The *BMD is composed cone by cone in reverse index order. A network cut is placed at the primary outputs as in [19] and the *BMD is intialized with the encoding function for the output word. In the succeeding steps, the cut frontier is always advanced only within a single cone until that cone has been fully traversed. Within this cone, the cut frontier is moved across fanout-free regions. At any time in the traversal, the next FFR to be placed behind the cut is the one with the heighest weight. The gate functions within an FFR that has moved behind the cut frontier are composed into the *BMD in reverse topological order.

Figure 3.18 shows the weights of the FFRs and the movement of the cut frontier within cone 3 of Fig. 3.16. The individual cut frontiers are indicated by their endpoints which are marked by ascending numbers. Note that the cut frontier continues in the adjacent cones but is fixed there until the current cone is fully traversed.

The goal of the heuristics in this approach is to keep the intermediate *BMD sizes as small as possible. Performing the composition conewise aims at obtaining pseudo-Boolean functions of arithmetic nature after every completed cone. Advancing the cut frontier over fanout-free regions keeps the intermediate cut size small and has a high chance of composing arithmetic units such as bitwise sum and carry functions into the *BMD.

Table 3.2 and Table 3.3 show experimental results for the construction of *BMDs using Hamaguchi's method and the cone-based approach for a number of multiplier circuits with different word sizes. For each method, the maximum intermediate *BMD size is reported. Note that this number is also indicative of the runtime performance of the algorithms. As opposed to ROBDDs, *BMD operations such as *apply* and *compose* have possibly exponential worst-case complexities so that processing *BMDs with large numbers of nodes leads to long CPU times. As can be seen from the tables, the cone-based approach considerably reduces intermediate *BMD sizes.

Table 3.2 shows the results for multipliers generated by a self-written generator (CSA array architecture). Table 3.3 shows the results for multipliers which have been created using a commercial CAD system (Synopsis Design Compiler). The circuit name prefixes *dw_csa* and *dw_nbw* refer to CSA array and Wallace tree architectures, respectively.

For self-generated multipliers (Table 3.2), the maximum size of the *BMDs is 25% less on the average and 3% less in the worst case for the cone-based approach when compared to Hamaguchi's method [19]. The table shows that Hamaguchi's method works very well for multipliers

circuit name	bit vector widths X	Y	Z	Hamaguchi max. *BMD size	cone-based max. *BMD size	size reduction (%)
mult_4x4	4	4	8	126	84	33
mult_8x8	8	8	16	249	154	38
mult_16x16	16	16	32	482	342	29
mult_24x24	24	24	48	772	594	23
mult_32x32	32	32	64	1094	910	16
mult_48x48	48	48	96	1846	1734	6
mult_64x64	64	64	128	2719	2628	3

Table 3.2. Experimental results for multipliers in CSA array architecture produced by a self-written generator.

circuit name	bit vector widths X	Y	Z	Hamaguchi max. *BMD size	cone-based max. *BMD size	size reduction (%)
dw_nbw_4x4	4	4	8	229	82	64
dw_csa_4x4	4	4	8	1139	299	73
dw_nbw_4x6	4	6	10	282	130	53
dw_csa_4x6	4	6	10	19448	2345	87
dw_nbw_5x8	5	8	13	304	155	49
dw_csa_5x8	5	8	13	20929	693	96
dw_nbw_6x8	6	8	14	528	198	62
dw_csa_6x8	6	8	14	—	3428	—
dw_nbw_6x9	6	9	15	384	180	53
dw_csa_6x9	6	9	15	—	7839	—
dw_nbw_8x8	8	8	16	547	345	36
dw_csa_8x8	8	8	16	—	816369	—
dw_nbw_6x12	6	12	18	611	297	51
dw_csa_6x12	6	12	18	—	3985	—
dw_nbw_8x16	8	16	24	1642	623	62
dw_csa_8x16	8	16	24	—	41708	—

Table 3.3. Experimental results for multipliers produced by a commercial CAD system

that have a very regular structure. The self-generated circuits are based on a simple standard "text-book" multiplication scheme. With increasing bit widths, the advantage of the cone-based approach diminuishes, however, it is never worse than the simple backward traversal technique.

For industrial multipliers (see Table 3.3), however, the cone-based method yields a much larger size improvement of 75% on the average. For some circuits in CSA array architecture, Hamaguchi's method was not able to construct a *BMD due to exponential blow-up, whereas the cone-based approach succeeded. Note the extreme difference in performance when comparing the results for industrial multipliers with those for the very regular circuits produced by our own generator. For industrial multiplier circuits larger than the ones shown in Table 3.3, also the cone-based backward traversal was not able to compute the *BMD within reasonable amounts of CPU time.

Although exploiting additional structural information significantly improves *BMD synthesis for industrial multipliers, only small to medium sized multipliers can be handled. Practical multipliers as generated from commercial synthesis tools contain many local optimizations, e.g., specialized adder structures, that improve the performance of the circuit but on the other hand severely degrade performance of the *BMD synthesis process. Sometimes even local circuit transformations by merging of gates or rearranging operands to internal circuit functions can cause the *BMD synthesis to become too complex.

As an experiment, we performed local circuit transformations on industrial multiplier netlists with the goal of reversing the optimizations causing the problems. We used a technique to extract XOR functions and to identify partial product bits similar to the arithmetic bit-level extraction technique to be described in Section 5. In addition, we employed a set of rule-based circuit transformations. The transformation rules were derived from a manual analysis of the multiplier netlists. The transformations include identification of full-adder carry functions as well as substitution of equivalent signals, both based on functional analysis. The goal of these transformations is to make the multiplier netlist more regular so that the cone-based backward traversal approach becomes successful.

An example for such transformations is shown in Figure 3.19. The circuitry on the left-hand side has been transformed into the circuitry on the right-hand side. The gates in the transformed circuit are clustered into arithmetic bit-level entities such as partial product bits, XORs of adders and carry functions.

Table 3.4 shows the results of *BMD synthesis after applying circuit transformations for reversing local optimizations for a number of

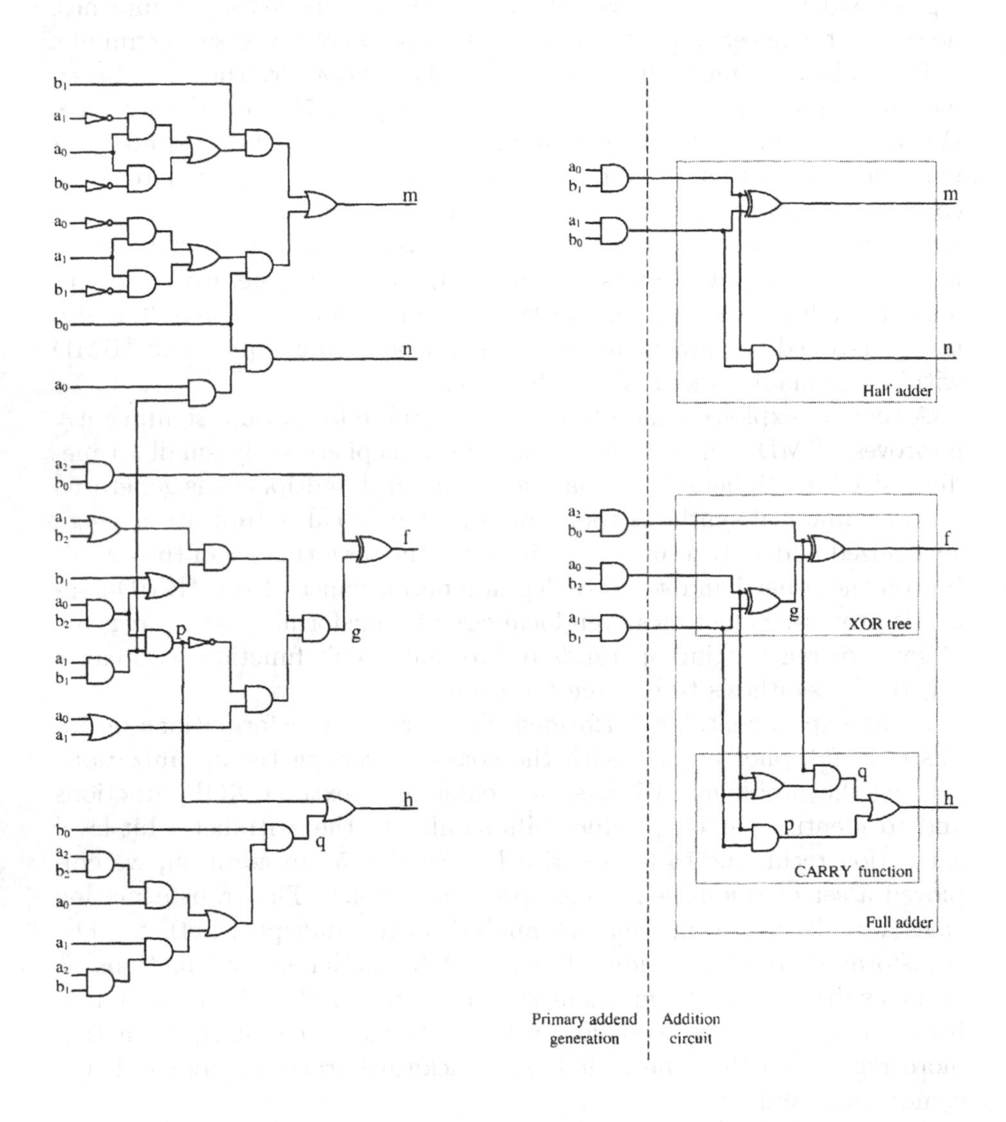

Figure 3.19. Example of local transformations for extracting arithmetic bit-level information

circuit name	bit vector widths X	Y	Z	Maximum *BMD size
dw_nbw_8x8	8	8	16	147
dw_csa_8x8	8	8	16	736
dw_nbw_16x16	16	16	32	1866
dw_csa_16x16	16	16	32	693
dw_nbw_16x26	16	26	42	1256
dw_csa_16x26	16	26	42	571

Table 3.4. *BMD synthesis results for some restructured multipliers

industrial multipliers. All multipliers have been generated by a commercial CAD system. The circuit denoted by prefix *dw_csa* and *dw_nbw* are multipliers in CSA array and Wallace tree architecture, respectively. For these circuits, we constructed the *BMDs within short CPU times. Note that a simple backward traversal as in Hamaguchi's method [19] fails for all circuits except *dw_nbw_8x8*.

These experiments show that rule-based transformations introducing regularity into the netlists are very helpful for constructing *BMDs. However, the transformation rules may work for one type of multiplier but fail for another. Clearly, the described approach is not a robust and general solution to the problem of *BMD synthesis from gate netlists. The intent of these experiments, however, was to demonstrate that success or failure of *BMD synthesis from gate netlists depend crucially on knowledge about the internal structure of the circuit and its subcomponents. It is worthwhile to investigate whether a more systematic approach to the extraction of such structural information can be developed. A technique of this kind is described in the following section.

5. Arithmetic Bit-Level Verification

Word-level decision diagrams such as *BMDs are very compact representations for arithmetic circuit functions that are also easy to manipulate. However, as shown in the previous section, constructing *BMDs from gate netlists so that they can be used in equivalence checking is difficult. Solutions to this problem which are robust enough for practical application do not yet exist. The *BMD synthesis procedures could benefit very much from additional structural information about the circuit. If the gate netlist could be decomposed into its sub-components, constructing a *BMD becomes feasible. However, as we will see in this section, if a representation in terms of sub-components is available, the verification problem itself becomes much easier and may be solved directly on this new representation.

In this section we describe a new approach that extracts such information about the arithmetic sub-components of a circuit. In some sense, it can be understood as a *reverse engineering* process. As described in the introduction (Section 1), most arithmetic circuits are constructed in a two-stage fashion as shown in Figure 3.2, (repeated in Figure 3.20 for convenience). The first stage computes primary addends, e.g., the partial product bits of a multiplier. The second stage adds these addends to produce the final result. Not only multipliers but also more general arithmetic expressions are implemented in this way (Figure 3.3). It is, however, not sufficient to simply identify the specific architecture

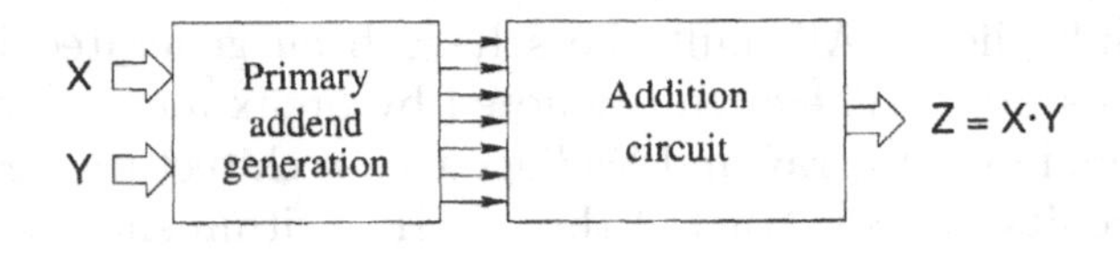

Figure 3.20. Basic multiplier structure

implemented in each of the two stages of an arithmetic circuit. For a successful *BMD synthesis, the bit-level building blocks such as the half and full adders in the addition circuit must be determined.

Also, one could think of employing such reverse engineering as a preprocessing step to a conventional cutpoint-based equivalence check. Noting that the number of possible architectures for the addition circuit is limited one could think of incorporating a complete set of architectures in the verifier frontend. The gate netlist would then be repeatedly compared against specifications in each of the provided architectures using a cutpoint-based equivalence checker. Unfortunately, the naive assumption that for the correct architecture there will be enough cutpoints for the checker to succeed is wrong. Even within one and the same architecture, e.g., a carry-save adder (CSA) array, there can be numerous implementation styles that exhibit hardly any similarity in terms of internal equivalences. As an illustration consider Figure 3.21 showing four ways of multiplying two decimal numbers.

All four cases can be implemented by the same architectures but have no internal equivalences at all. The adder stage of each row computes the *accumulated* sum of the previous rows. The accumulated sum values are different in all four variations. We experimentally verified the ab-

167 · 239	167 · 239	239 · 167	239 · 167
334	1503	239	1673
501	501	1434	1434
1503	334	1673	239
39913	39913	39913	39913

Figure 3.21. Multiplication example (decimal numbers)

sence of internal equivalences using the 16x16 bit multiplier C6288. We modified the circuit by swapping its operands. Since multiplication is commutative C6288 with swapped operands must be equivalent to the original version. Proving this by our equivalence checker [27], however, turned out to be impossible. All internal equivalences were lost, except for the ones belonging to the partial products in the first circuit level.

As can be seen, regardless of the intended application — *BMD synthesis or standard equivalence checking — reverse engineering at the architecture level is not sufficient. It is necessary to target building blocks at lower levels of abstraction. A first attempt in this direction was made in [28]. This work explored mapping the gates in a multiplier netlist to a set of component cells by pattern matching using a logic programming tool. However, the architectures of the cells have to be known a priori and have to be given to the tool along with the circuit to be verified. In this section we also describe a technique operating directly on the gate level. This technique, however, decomposes the gate netlist of an arithmetic circuit into its *smallest arithmetic units.* Instead of identifying word-level operations as a whole these are broken down into arithmetic operators on single-bit signals. The output of the extraction procedure is an *arithmetic bit level* description of the circuit. Addition at this level is reduced to addition modulo 2 and generation of carry signals. In general terms, the proposed approach can be summarized as follows:

1 Decompose the two combinational circuits – where possible – into networks of 1-bit addition primitives, such as XOR, half adder, full adder (arithmetic bit level).

2 Prove equivalence of corresponding circuit outputs on the arithmetic bit level using commutative and associative laws.

The arithmetic bit level representation of a circuit is very desirable. It could, for example, serve as a preprocessing step to a *BMD synthesis procedure or could be used by the frontend of a standard equivalence checker to produce an easy-to-verify specification of the arithmetic circuit. However, given such a representation the verification task becomes so easy that checking circuit equivalence can be solved without any additional techniques.

5.1 Verification at the Arithmetic Bit Level

Any combinational circuit which performs the addition of binary bit vectors such as the addition stage in a multiplier can be represented as a composition of half and full adders. Figure 3.22 shows the gate schematics of a half adder. In the sequel, we will use the half adder symbol shown on the right side of Figure 3.22.

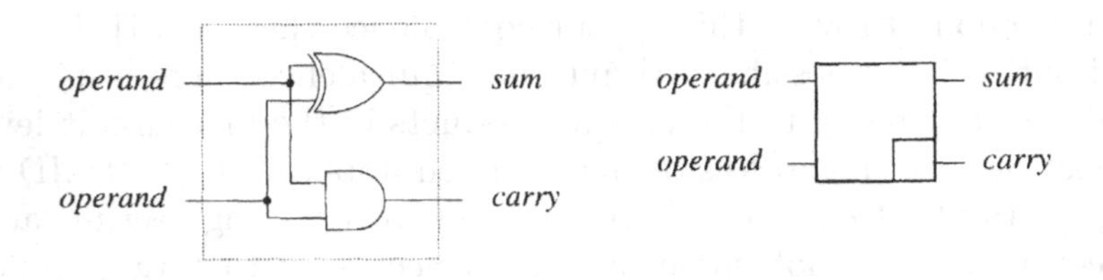

Figure 3.22. Half adder, schematics and symbol

A full adder can be completely decomposed into half adders. We make use of this fact in our choice of arithmetic bit level representation. Figure 3.23 shows a possible implementation of a full adder and the corresponding network composed of three half adders P, Q, R. Half adder R adds the two carry bits c_1 and c_2 of the half adders P and Q and produces the full adder carry output w. Note that because the two signals c_1 and c_2 can never assume the logic value 1 simultaneously, the carry output of R produces a constant 0.

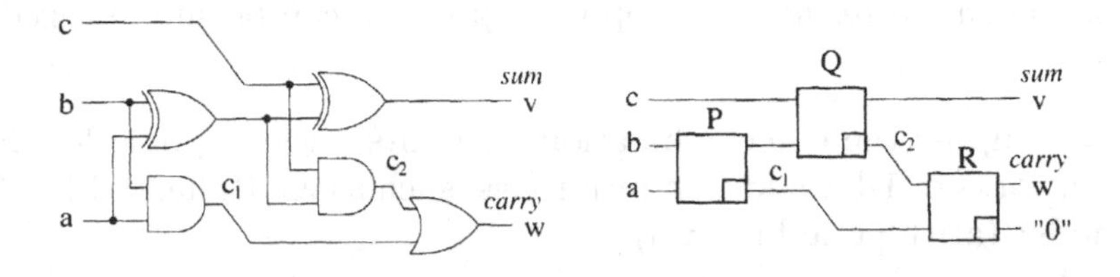

Figure 3.23. Full adder decomposed into half adders

Once we have obtained a representation of an addition circuit that is only composed of half adders, we speak of a *half adder network* or the *arithmetic bit level representation* of the circuit. This representation allows for a very efficient equivalence checking procedure. In the following, we introduce a mathematical model for the arithmetic bit level and develop the theoretical background of our verification procedure.

Definition 3.1 *An* addition graph *is a triple* $(G(V,E), R, F)$. $G(V,E)$ *is a bipartite directed graph with vertex set* V *and directed edge set* E. *The vertex set* V *consists of three disjoint subsets,* $V = S \cup C \cup I$. *The vertices in* S *have exactly two immediate predecessors, and are called* sum nodes. *The vertices in* I *have no predecessors and are called* primary addends. *The vertices in* C *have no predecessors and are called* carry nodes.

R is a relation, $R \subseteq (C \times S)$ *and* F *is a set of Boolean functions.*

The addition graph is associated with a half adder network as follows. Each sum node is associated with the sum output of a half adder in the network. Each carry node is associated with the carry output of a half adder. Each primary addend is associated with an input of the half adder network.

Two vertices v *and* w *are connected by a directed edge* (v, w), *if the half adder associated with* w *has the signal associated with* v *as operand.*

For $c \in C$ *and* $s \in S$ *it is* $(c, s) \in R$ *if and only if* c *and* s *are associated with the output signals of the same half adder in the network.*

With each vertex $v \in V$ *we associate the Boolean function* $f_v \in F$ *in terms of the primary addends that is implemented by the signal corresponding to* v *in the half adder network.*

For illustration of this definition, Figure 3.24 shows the addition graph of the full adder of Figure 3.23. Note that the primary addends and the

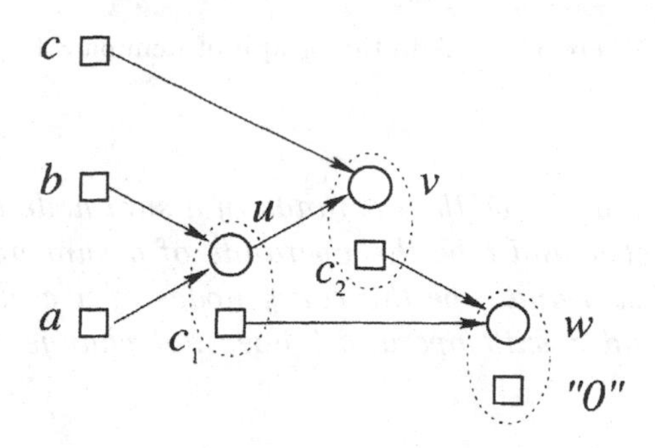

Figure 3.24. Addition graph for full adder

carry nodes are the source nodes of an addition graph, and are also referred to as *addends* in the following. In Figure 3.24, addends are represented by boxes, sum nodes are represented by circles. The relation between carry and sum nodes is indicated by dashed lines. Nodes v and w are sinks of the addition graph and correspond to outputs v and w of the half adder network.

The modelling of a half adder by two separate nodes in the addition graph may seem awkward. Note, however, that our definition leads to a decomposition of the half adder network into graph entities such that all but the source vertices correspond to XOR operations. Therefore, each sum node in the graph can be associated with the sum modulo 2 of all source nodes in its transitive fanin. This facilitates the manipulation of the graph structure.

In the following, without loss of generality, we assume that the addition graph is a forest of *trees*. If the addition graph obtained from the original half adder network does not have tree structure, we can always generate a forest of trees by duplication of appropriate graph portions including primary addends.

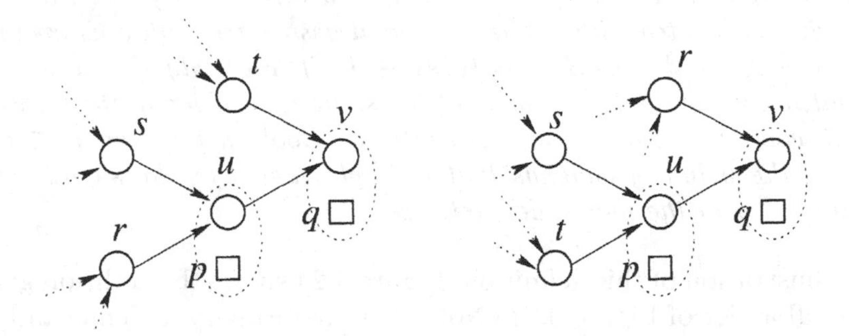

Figure 3.25. Addition graph of Lemma 3.1

Lemma 3.1 *Let r and s be the operands of a sum node u in an addition graph. Further, let u and t be the operands of a sum node v, as shown in Figure 3.25. Let p and q be the carry nodes of u and v, respectively. Exchanging operand r with operand t does not change f_v and does not change $f_p \oplus f_q$.*

Proof 3.1 *Function f_v does not change because addition modulo 2 is commutative. The function $f_p \oplus f_q$ does not change, because $(r \cdot s) \oplus ((r \oplus s) \cdot t) = (t \cdot s) \oplus ((t \oplus s) \cdot r)$.*

Half adder networks implementing practical addition stages have the special property that each addition tree computes a digit of a binary encoded integer. The carry signals of the addition tree for digit i all feed into the addition tree for the next digit, $i + 1$. This can be exploited when checking the equivalence of addition trees in practical addition networks.

Lemma 3.2 *The output functions of two addition trees T and $\tilde{T}$ (Figure 3.26) are equivalent if the following conditions are true.*

1. *The sets of primary addends for T and $\tilde{T}$ are identical ($I_T = I_{\tilde{T}}$).*
2. *There exists an addition tree S such that the set of all carry nodes being addends for T is identical with the set of carries generated in S. The same holds for $\tilde{T}$ and some addition tree $\tilde{S}$.*
3. *The output functions of S and $\tilde{S}$ are equivalent.*

Proof 3.2 *If the output functions of S and $\tilde{S}$ are equivalent, then the sum modulo 2 of all carries generated in S is equivalent to the sum modulo 2 of all carries generated in $\tilde{S}$. This follows from the observation that S can be transformed into $\tilde{S}$ by a sequence of operand swaps according to Lemma 3.1. T as well as $\tilde{T}$ compute the modulo 2 sum of the primary addends and the carries of S.*

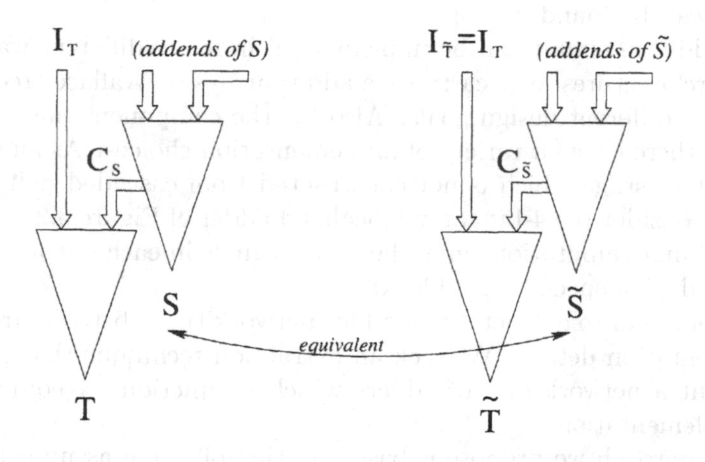

Figure 3.26. Illustration of Lemma 3.2

Once we have a representation of an addition circuit as a half adder network, the equivalence check using Lemma 3.2 is straightforward. Note that finding addition tree S for addition tree T in condition 2 is trivial in practice, since S is located in the immediate structural vicinity of T. The correspondences $\tilde{S}$ with S and $\tilde{T}$ with T are known from the given equivalence checking task.

Note the recursive nature of Lemma 3.2: the equivalence of the output digit i (tree T) depends on the equivalence of digit $i-1$ (tree S). The terminal case of the recursion is digit 0 where no carry-ins exist and only condition 1 of the lemma needs to be checked. The total run-time of the equivalence check according to Lemma 3.2 is linear in the number of half adders which is proportional to circuit size.

Another possibility to verify addition circuits on the arithmetic bit level is to manipulate the circuits using the operation of Lemma 3.1 until both circuits have the same structure and contain enough internal equivalences for a standard equivalence checking procedure to be successful.

The problem that remains to be solved, however, is how to extract the arithmetic bit level representation from the gate netlist of an addition circuit. This is subject of the following section.

5.2 Extracting the Half Adder Network

In the following, we give an intuitive description of the techniques used for extracting an arithmetic bit-level representation of a circuit in the form of a half adder network. A more detailed treatment of this subject can be found in [29].

An addition circuit can be implemented in many different ways. Different architectures, e.g. carry-save adder arrays or Wallace trees, exist, aiming at different design goals. Also for the components and subcomponents there exists a variety of implementation choices. As an example of an adder stage which is not constructed from cascaded half and full adders, consider the 4-bit carry-lookahead adder of Figure 3.27. In order to speed up computation time, the carry signals in each output cone are generated by a special logic block.

It is our goal to extract a half adder network that abstracts from such implementation details. We seek an extraction technique that produces as output a network of half adders which is functionally equivalent to the implementation.

The approach we propose is based on the following assumption: The predominant operation at the bit level is the computation of exclusive OR. This logic function is part of every implementation of binary addition. We use Boolean reasoning techniques [27] to detect XOR relationships in the original circuit. Note that there are many possibilities to

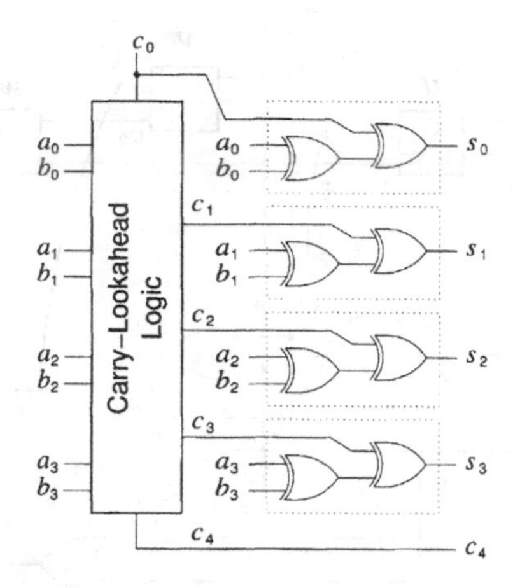

Figure 3.27. 4-bit carry-lookahead adder

implement XOR detection algorithms, e.g. using SAT, local BDDs or structural hashing techniques. In order to trade off performance against quality of results, it is desirable to have several phases with different algorithms. Although we have not experimented with this, we believe that as a first extraction phase, a structure-based functional hashing technique, e.g., the technique based on AND/INVERTER graphs of [30], could be very efficient to extract the majority of the XOR functions in an arithmetic circuit. The few remaining XOR functions could then be detected by a more powerful yet more time-consuming functional analysis based on SAT, BDDs or ATPG.

Guided by the detected XORs we construct a network of half adders as a reference circuit. We store implications between nodes in the original circuit and the half adder network. The stored implications establish a mapping between the nodes of the original and the reference circuit.

As an example, consider the implementation of a full adder shown in Figure 3.28.

Using Boolean reasoning techniques it is possible to prove that the signal x can be expressed as the exclusive OR of signals a and b. As a consequence, in the reference circuit, we insert a half adder node u with operands a and b and store implications reflecting the equivalence of the sum output of the half adder and node x. Also, signal p can be expressed as the exclusive OR of x and c. We insert a half adder node v with operands x and c and store the equivalence of the sum output with signal p.

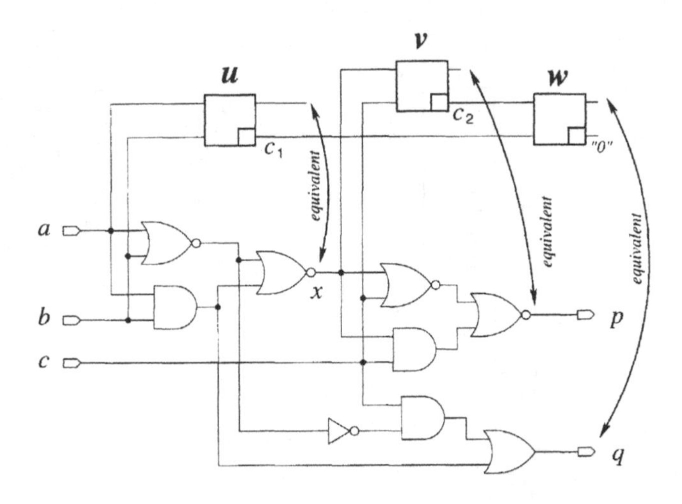

Figure 3.28. Full adder implementation and mapped half adder network

Now that the half adders u and v exist, it is possible to express signal q as an exclusive OR of the carry outputs c_1 of u and c_2 of v. Also, we can identify the implication $c_1 = 1 \rightarrow c_2 = 0$ which is equivalent to $c_1 \cdot c_2 = 0$, for all possible input vectors of the adder circuit. Therefore we insert half adder w with operands c_1 and c_2, and we store the information that the carry output of this half adder produces a constant 0. We also store an equivalence pointer between the sum output of w and the output q of the adder circuit. We now have a complete mapping of the adder circuit as a half adder network.

Note that although function q implements the majority function, $q = (a+b)c + ab = ab + ac + bc$, of the inputs a, b, c and not an XOR function of any of these operands, we can still find a mapping for this node by using signals from the reference circuit.

When detecting an XOR relationship of the form $y = a \oplus b$ for some signal y in the original circuit, with a and b being signals in the original or in the reference circuit, it is actually not sufficient to insert a half adder with operands a and b. It could be that an operand has to be inverted in order to make the half adder useful as an operand later. Since the correct operand phases cannot be determined by the XOR detection ($y = a \oplus b = \overline{a} \oplus \overline{b}$), we add not only one half adder for each XOR found but all four half adders corresponding to the four possible combinations of inversions of the operands.

The Boolean analysis underlying this procedure is local and of fairly low complexity. An efficient implementation can be based on ATPG, SAT, decision diagrams or structural hashing [3].

5.3 Verification Framework

The proposed approach can be added as an additional heuristic to existing equivalence checking frameworks. Equivalence checking is run for given circuits in the usual way until standard techniques abort by lack of internal equivalences. If there are large regions without internal equivalences, the extraction procedure of Section 5.2 is activated, attempting to generate an arithmetic bit level representation of the pathological region. This can be successful, if the region is indeed an arithmetic block. If the circuit contains a multiplier, standard equivalence checking will be successful in identifying internal equivalences for many nodes in the circuit, including the partial products of the multiplier. However, it will fail to process the subsequent addition circuit. After extracting the arithmetic bit level representation the verification can be completed.

Commercial equivalence checkers offer practical solutions for verifying circuits containing multipliers by *black-boxing* these multipliers. If, however, arithmetic expressions have been optimized as in Figure 3.3, the individual multiplication operators can no longer be identified. It is interesting to note that the arithmetic bit-level extraction approach proposed here is insensitive to such expression optimizations, since the general two-stage structure of primary addend generation followed by an addition circuit still prevails.

The proposed extraction procedure will fail to extract an arithmetic bit level description if the multiplier circuit contains an error. This, however, is easily detected by a simulation step earlier in the verification flow. As stated previously, multipliers are highly random-pattern testable so that a buggy design is usually detected by only a small number of random patterns.

5.4 Experimental Results

The described techniques have been implemented as a part of the HANNIBAL [27] tool. Table 3.5 shows some of our results for extracting the half adder networks for multipliers of different origin, bit widths and architectures. In our experiments we found that for multipliers generated by our own self-written generator tools it was much easier to extract the arithmetic bit-level representation than for circuits generated by commercial synthesis tools. These circuits are much more "sophisticated" than their academic counterparts which are constructed using common text-book components and architectures. For reasons of space, we do not report verification results for circuits generated by our own tools.

The first column shows the origin of the circuit. We have experimented with multipliers generated by two different commercial synthesis tools, one of them being Synopsys Design Compiler ™(labeled "comm. 1") in Table 3.5. Additionally, we have three different versions of C6288 from the well-known ISCAS-85 benchmark set in the table. Circuit *c6288* is the original circuit, circuit *c6288nr* is its non-redundant version, and circuit *c6288opt* is the result of optimizing *c6288* using SIS with *script.rugged.*

The various multipliers process signed or unsigned numbers (column 2). Some benchmarks are based on Booth encoding (marked "b" in column 3), others are not (marked "n" in column 3). Columns 4 to 6 show the bit widths of multiplication operands, X, Y, and result, Z. Column 7 shows the circuit size given as the number of connections in the netlist. The last column reports on the CPU time on a 1300 MHz PC running Linux.

The benchmarks differ greatly with respect to the architectures used, such as Wallace trees and arrays of RCA or CSA adders. For all these architectures, the arithmetic bit level could be extracted within short CPU times. Note that due to the Boolean nature of our extraction technique the arithmetic bit level can also be obtained if the multiplier has been been optimized using standard logic synthesis techniques. This is illustrated by means of *c6288opt* and logic synthesis by SIS.

We verified the equivalence between any pair of multipliers with the same operand widths and number interpretation (signed/unsigned) using the equivalence check of Lemma 3.2. After the arithmetic bit level was extracted, the actual equivalence check in all cases took only a fraction of a second.

In a second experiment (Table 3.6) we ran the arithmetic bit level extraction algorithm on circuits computing larger arithmetic expressions containing several multipliers. These netlists were generated by Synopsys DC Ultra ™, which features advanced arithmetic optimization capabilities as described in Section 5.1 by identifying and merging arithmetic addition trees.

Circuit *MAC1* computes the expression `y = a0*b0 + a1*b1`, and circuit *MAC8* computes `y = a0*b0 + ...+ a8*b8`. Column 2 of Table 3.6 shows the number of multipliers in each expression. The remaining columns report the bit width of the result y, the size of each circuit given as the number of connections and the CPU time needed to extract the arithmetic bit level representation. Note that the multiplication op-

circuit origin	u/s	Booth enc.	bit vector widths X	Y	Z	size (# conns.)	CPU time (h:mm:ss)
comm. 1	s	b	22	6	28	1393	0:00:08
comm. 1	s	n	22	6	28	1372	0:00:22
comm. 1	s	n	6	22	28	1394	0:00:27
comm. 1	u	b	22	6	28	1424	0:00:07
comm. 1	u	b	6	22	28	1326	0:00:09
comm. 1	u	n	22	6	28	1254	0:00:08
comm. 1	u	n	6	22	28	1499	0:00:10
comm. 1	u	n	22	6	28	1362	0:00:11
c6288	u	n	16	16	32	5568	0:00:33
c6288nr	u	n	16	16	32	4698	0:01:37
c6288opt	u	n	16	16	32	4721	0:00:21
comm. 1	s	b	16	16	32	2324	0:00:29
comm. 1	u	n	16	16	32	2682	0:00:44
comm. 1	u	n	16	16	32	2778	0:01:37
comm. 2	s	b	16	16	32	2340	0:00:26
comm. 2	u	b	16	16	32	2443	0:00:20
comm. 2	u	n	16	16	32	2740	0:00:38
comm. 2	u	n	16	16	32	2336	0:00:22
comm. 1	s	b	15	22	37	2955	0:00:42
comm. 1	s	b	22	15	37	3052	0:00:54
comm. 1	s	n	15	22	37	3652	0:02:14
comm. 1	u	b	15	22	37	2944	0:00:49
comm. 1	u	b	22	15	37	3174	0:00:53
comm. 1	u	n	15	22	37	3541	0:01:23
comm. 1	u	n	22	15	37	3524	0:01:14
comm. 1	u	n	15	22	37	3652	0:01:54
comm. 2	s	b	24	24	48	5178	0:01:32
comm. 2	u	b	24	24	48	5319	0:01:27
comm. 2	u	n	24	24	48	6410	0:03:40
comm. 2	u	n	24	24	48	5424	0:01:49
comm. 2	s	n	32	32	64	11636	0:24:13
comm. 2	u	b	32	32	64	9265	0:04:34
comm. 2	u	n	32	32	64	11612	0:13:35
comm. 2	u	n	32	32	64	9792	0:05:48
comm. 2	s	b	48	48	96	20203	0:29:50
comm. 2	u	b	48	48	96	20464	0:22:09
comm. 2	u	n	48	48	96	22368	0:29:21

Table 3.5. Arithmetic bit level extraction on multiplier gate netlists

circuit	# mult.'s	result bit width	# conn.'s	# CPU time (s)
MAC1	2	17	1528	7
MAC8	9	17	12034	95

Table 3.6. Arithmetic bit level extraction on expressions

erations cannot be isolated as blocks in the gate netlist due to the arithmetic optimization by the synthesis tool. The partial product bits of each sub-expression are added by one optimized addition tree which can be extracted using the proposed algorithm. Both circuits were proven to be equivalent to their specification using Lemma 3.2.

6. Conclusion

In this chapter we have summarized the results of over a decade of world-wide research efforts to improve on the verification of arithmetic circuits. Even after so many years, equivalence checking of arithmetics, especially multipliers, is still a very hard problem for which only partial solutions exist.

Verification techniques that use specific properties of the implemented arithmetic functions (Section 2) work only in special cases where such properties can be identified [7], or they are not robust in the sense that they cannot sufficiently handle certain architectures [8, 9].

Solutions based on bit level decision diagrams (Section 3) such as [4, 11] suffer from high complexity and may lack robustness, even if the BDDs are not built for the circuit outputs directly but certain structural properties of the arithmetic circuits (e.g. "structural dependence" [11]) are exploited.

The greatest advances have been achieved with the invention of *word-level decision diagrams* (Section 4). Among those, *BMDs [18] have the greatest promise because they can efficiently represent integer multiplication. However, they require word-level information about a design which is often not available and difficult to extract from a given bit level implementation. If *BMDs are to be constructed from a gate netlist, so far, only a backwards traversal synthesis technique can be successful, and it requires fine-tuned heuristics to select the right cut frontiers. Even then, for many practical circuits as generated from commercial synthesis tools, *BMDs cannot be constructed because of excessive consumption of computation ressources.

Finally, we have presented a pragmatic approach for equivalence checking of arithmetic circuits including multipliers (Section 5). The method is based on a bit level reverse-engineering approach. The main challenge is to efficiently extract an arithmetic bit level description of a circuit from a given gate netlist. As described in Section 5.2, the presented technique is not general, i.e., in practice it relies on certain assumptions about the internal structure of the circuits to be verified. Fortunately, these assumptions are often fulfilled for today's synthesis enginges. In a practical evaluation, the method has been tested on various multipliers and other arithmetic circuits and proved very promising. The approach can easily be integrated into standard equivalence checking frameworks and can increase the robustness of conventional equivalence checkers for arithmetic circuits.

7. Future Perspectives

Formal verification of arithmetic circuits remains to stay an interesting field of research with many open questions, not only in equivalence checking. To a growing extent, formal verification by equivalence checking in industrial design flows is being complemented by property checking. Especially since the advent of *Bounded Model Checking* (BMC) [31] a few years ago, property checkers became capable of verifying designs of realistic sizes. Bounded model checkers are based on purely combinational models of the circuits to be verified. Also in property checking, arithmetic circuits pose problems to the verification tools. The situation becomes particularly difficult if the arithmetic expressions are intertwined with Boolean control logic so that pure word-level representations are not appropriate.

It is not at all obvious how to cope with arithmetic circuitry in such situations. There may be cases where a word-level diagram is a desirable form of representation of a circuit. However, as demonstrated in this chapter, even after many years of research it is not clear how to obtain word-level diagrams from gate or bit-level circuit representations. One possibility which is currently being explored in our research group is using an intermediate arithmetic bit-level representation as presented in Section 5 also in property checking. Many other possibilities exist that need to be explored. For example, SAT-based techniques as used in BMC have shown to be very effective for verifying control-dominated designs. On the other hand, they behave poorly on datapath circuits. Different solving techniques, e.g., such as ILP (*integer linear programming*), have proven to be more successful on pure arithmetic problems. However, typical property checking tasks contain expressions over both, datapath and control logic signals. Handling the control parts of a problem efficiently such that the power of ILP solver on the datapath parts can be exploited is a difficult and open problem with many research opportunities [32, 33]. It may also be worthwhile to consider other solving techniques such as CLP (*constraint logic programming*), e.g. along the lines of [34].

As has been successful for equivalence checking, the goal for the development of efficient property checkers will be a multi-engine approach with each engine having its specific strengths and weaknesses. An effective set of heuristics implemented in the verifier front-end will have to pre-process each verification problem in order to identify adequate sub-problems which can be solved by the appropriate solving engines. A great amount of research is necessary to develop these new heuristics and new solving engines and to determine the interactions between all components.

References

[1] W. Kunz, "An efficient tool for logic verification based on recursive learning," in *Proc. International Conference on Computer-Aided Design (ICCAD-93)*, pp. 538–543, Nov. 1993.

[2] D. Brand, "Verification of large synthesized designs," in *Proc. International Conference on Computer-Aided Design (ICCAD-93)*, pp. 534–537, 1993.

[3] A. Kühlmann and F. Krohm, "Equivalence checking using cuts and heaps," in *Proc. Design Automation Conference (DAC-97)*, pp. 263–268, June 1997.

[4] J. R. Bitner, J. Jain, M. S. Abadir, J. A. Abraham, and D. S. Fussell, "Efficient algorithmic circuit verification using indexed BDDs," in *Proc. Fault Tolerant Computing Symposium (FTCS-94)*, pp. 266–275, 1994.

[5] J. Jain, R. Mukherjee, and M. Fujita, "Advanced verification techniques based on learning," in *Proc. 32nd ACM/IEEE Design Automation Conference (DAC-95)*, pp. 420–426, June 1995.

[6] Y. Matsunaga, "An efficient equivalence checker for combinational circuits," in *Proc. Design Automation Conference (DAC-96)*, pp. 629–634, June 1996.

[7] M. Fujita, "Verification of arithmetic circuits by comparing two similar circuits," in *Proc. International Conference on Computer Aided Verification (CAV '96)* (R. Alur and T. A. Henzinger, eds.), no. 1102

in Lecture Notes in Computer Science, pp. 159–168, Springer, August 1996.

[8] Y.-T. Chang and K.-T. Cheng, "Induction-based gate-level verification of multipliers," in *Proc. International Conference on Computer-Aided Design (ICCAD-01)*, (San Jose, CA), pp. 190–193, 2001.

[9] Y.-T. Chang and K.-T. Cheng, "Self-referential verification of gate-level implementations of arithmetic circuits," in *Proc. Design Automation Conference (DAC-02)*, pp. 311–316, 2002.

[10] C. S. Wallace, "A suggestion for a fast multiplier," *IEEE Transactions on Electronic Computers*, vol. EC-13, pp. 14–17, February 1964.

[11] T. Stanion, "Implicit verification of structurally dissimilar arithmetic circuits," in *Proc. International Conference on Computer Design (ICCD-99)*, pp. 46–50, October 1999.

[12] R. Bryant, "Graph-based algorithms for boolean function manipulation," *IEEE Transactions on Computers*, vol. 35, pp. 677–691, August 1986.

[13] J. Burch, "Using BDDs to verify multipliers," in *Proc. Design Automation Conference (DAC-91)*, pp. 408–412, 1991.

[14] Y. T. Lai and S. Sastry, "Edge-valued binary decision diagrams for multi-level hierarchical verification," in *Proc. Design Automation Conference (DAC-95)*, pp. 254–260, 1995.

[15] E. M. Clarke, M. Fujita, P. McGeer, K. L. McMillan, J. Yang, and X. Zhao, "Multi-terminal binary decision diagrams: an efficient data structure for matrix representation," in *Proc. International Workshop on Logic Synthesis*, pp. (P6a) 1–15, 1993.

[16] E. M. Clarke, K. L. McMillan, X. Zhao, M. Fujita, and J.-Y. Yang, "Spectral transforms for large boolean functions with application to technology mapping," in *Proc. 30th ACM/IEEE Design Automation Conference (DAC-93)*, (Dallas, TX), pp. 54–60, June 1993.

[17] R. I. Bahar, E. A. Frohm, C. M. Gaona, G. Hachtel, E. Macii, A. Pardo, and F. Somenzi, "Algebraic decision diagrams and their application," in *Proc. International Conference on Computer-Aided Design (ICCAD-93)*, pp. 188–191, 1993.

[18] R. Bryant and Y. A. Chen, "Verification of arithmetic functions by binary moment diagrams," in *Proc. Design Automation Conference (DAC-95)*, pp. 535–541, 1995.

[19] K. Hamaguchi, A. Morita, and S. Yajima, "Efficient construction of binary moment diagrams for verifying arithmetic circuits," in *Proc.*

International Conference on Computer-Aided Design (ICCAD-95), pp. 78–82, November 1995.

[20] R. Drechsler, B. Becker, and S. Ruppertz, "K*BMDs: A new data structure for verification," in *Proc. European Design & Test Conference*, pp. 2–8, 1996.

[21] U. Kebschull, E. Schubert, and W. Rostenstiel, "Multi-level logic based on functional decision diagrams," in *Proc. European Design Automation Conference (EDAC-92)*, pp. 43–47, 1992.

[22] R. Drechsler, B. Becker, A. Sarabi, M. Theobald, and M. Perkowski, "Efficient representation and manipulation of switching functions based on ordered kronecker functional decision diagrams," in *Proc. Design Automation Conference (DAC-94)*, pp. 415–419, 1994.

[23] Y.-A. Chen and R. E. Bryant, "*PHDD: An efficient graph representation for floating point circuit verification," in *Proc. International Conference on Computer-Aided Design (ICCAD-97)*, pp. 2–7, November 1997.

[24] M. Keim, M. Martin, B. Becker, R. Drechsler, and P. Molitor, "Polynomial formal verification of multipliers," in *VLSI Test Symposium*, pp. 150–155, 1997.

[25] Y.-A. Chen and J.-C. Chen, "Equivalence checking of integer multipliers," in *Proc. Asia and South Pacific Design Automation Conference (ASPDAC-01)*, (Yokohama, Japan), pp. 196–174, 2001.

[26] M. Abramovici, M. Breuer, and A. Friedman, *Digital Systems Testing and Testable Design.* Piscataway, New Jersey: IEEE Press, 1994.

[27] W. Kunz and D. Stoffel, *Reasoning in Boolean Networks - Logic Synthesis and Verification Using Testing Techniques.* Boston: Kluwer Academic Publishers, 1997.

[28] H. Simonis, "Formal verification of multipliers," in *Proceedings of the IFIP WG10.2 WG10.5 International Workshop on Applied Formal Methods for Correct VLSI Design* (L. J. Claesen, ed.), (North Holland), pp. 267–286, Elsevier Science Publishers B.V., 1990.

[29] D. Stoffel and W. Kunz, "Equivalence checking of arithmetic circuits on the arithmetic bit level," *submitted to IEEE Transactions on Computer-Aided Design*, 2003.

[30] A. Kuehlmann, M. K. Ganai, and V. Paruthi, "Circuit-based boolean reasoning," in *Proc. Design Automation Conference (DAC-01)*, pp. 232–237, June 2001.

[31] A. Biere, A. Cimatti, E. M. Clarke, M. Fujita, and Y. Zhu, "Symbolic model checking using SAT procedures instead of BDDs,"

in *Proc. International Design Automation Conference (DAC-99)*, pp. 317–320, June 1999.

[32] Z. Zeng, M. Ciesielski, and B. Rouzeyre, "Functional test generation using constraint logic programming," in *Proc. IFIP VLSI-SOC Conference*, December 2001.

[33] R. Brinkmann and R. Drechsler, "RTL-datapath verification using integer linear programming," in *VLSI Design*, pp. 741–746, 2002.

[34] Z. Zeng, M. Ciesielski, and B. Rouzeyre, "LPSAT: A unified approach to RTL satisfiability," in *Proc. Design, Automation and Test in Europe Conference (DATE-2001)*, pp. 398–402, March 2001.

[35] S. Kimura, "Residue BDD and its application to the verification of arithmetic circuits," in *Proc. Design Automation Conference (DAC-95)*, pp. 542–545, 1995.

[36] T. Kim, W. Jao, and S. Tjian, "Arithmetic optimization using carry-save-adders," in *Proc. Design Automation Conference (DAC-98)*, pp. 433–438, 1998.

Chapter 4

APPLICATION OF PROPERTY CHECKING AND UNDERLYING TECHNIQUES

Infineon's Circuit Verification Environment

Raik Brinkmann
Infineon Technologies AG
raik.brinkmann@infineon.com

Peer Johannsen
Infineon Technologies AG
peer.johannsen@infineon.com

Klaus Winkelmann
Infineon Technologies AG
klaus.winkelmann@infineon.com

Abstract This article gives an in-depth view of the use of formal property verification at Infineon Technologies AG. We present the method and tool from a user perspective, and also discuss some aspects of its underlying innovations. Finally we present a range of applications high-lighting the strong relevance of property checking for today's complex design projects.

Keywords: Property checking, SAT algorithm, circuit verification, CVE

Introduction

Infineon Technologies AG offers a broad range of semiconductor products for target markets such as mobile communications and networks,

R. Drechsler (ed.), Advanced Formal Verification, 125-166.

access control and network security, car electronics, and more. In order to meet its highly demanding cost and quality targets, Infineon is using an advanced design flow incorporating state-of-the-art commercial tools, as well as innovative in-house tools.

For guaranteeing functional correctness, formal property-checking has become an integral part of this design flow, and has been used in more than 20 projects to date, including large ASICs.

The focus of property checking is on the block level, i.e. for establishing functional correctness of design blocks, where a block is typically an object handled by a single designer, comprising a few hundred up to approx. 10000 lines of HDL code. It may consist itself of a hierarchy of VHDL entities, but seldom of more than two levels.

This focus on individual design blocks is motivated as follows: Local errors i.e. errors whose cause resides within a single block, constitute more than 50% of all errors. Often they are hard to localize from a systems view and delay ramp-up of system simulation. As the probability of correct functioning of a chip equals at most the product of the respective probabilities for all blocks, extreme quality blocks are key for predictable system quality.

For verifying the integration of formally-verified blocks into a larger entity, e.g. a complete ASIC, state-of-the-art simulation techniques are used, which are outside the scope of this paper.

This article gives an in-depth view of the use of formal property verification at Infineon. We present the approach not only from a user perspective, but also discuss some aspects of its underlying technology. In particular we discuss a novel approach of dealing with bit-vector operations, as well as the exploitation of symmetries to speed up proof performance.

Finally we present a range of applications high-lighting various use cases, and discuss relevant parameters such as project size, project phase when verification is applied, verification goals, verification effort, cost and quality.

1. Circuit Verification Environment: User's View

1.1 Tool Environment

The Infineon design flow includes the formal verification tool-set CVE (Circuit Verification Environment), which has been developed by Siemens and Infineon for more than a decade. CVE is mostly used by these two companies, but also made available to a number of large European electronics firms.

CVE consists of language front-ends including VHDL and Verilog, the gatecomp equivalence checker and a property checker called gateprop.

1.2 The gateprop Property Checker

Functional verification using gateprop, is based on

- compiling (automatically) the design into an internal finite state machine representation, and
- formalising (manually) its specification using a simple temporal property language, called ITL.
- The formalised specification together with the FSM is then transformed into a bounded model checking problem, which is checked for satisfiability.

This yields a work flow as outlined in Figure 4.1.

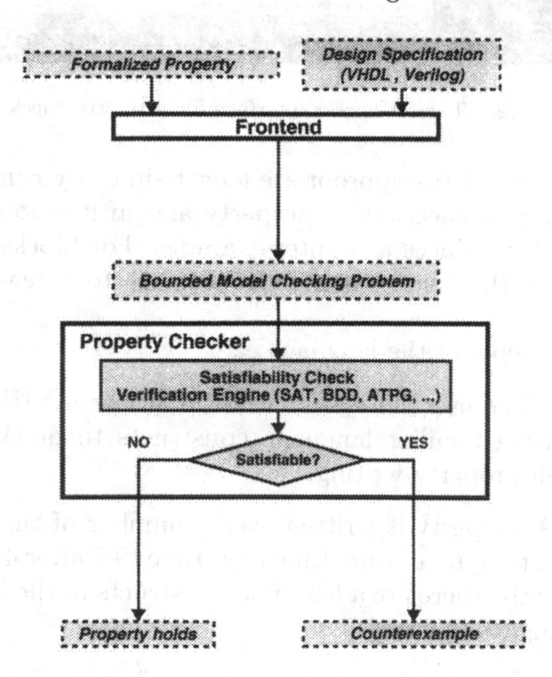

Figure 4.1. Property Checking Work Flow

Figure 4.2 shows a desk-top with the CVE graphical user interface for property checking in the foreground window, and an editor with a property file in the background.

An ITL property is essentially a constraint on the design's signals over a finite time interval. For the property to be valid means to hold for every

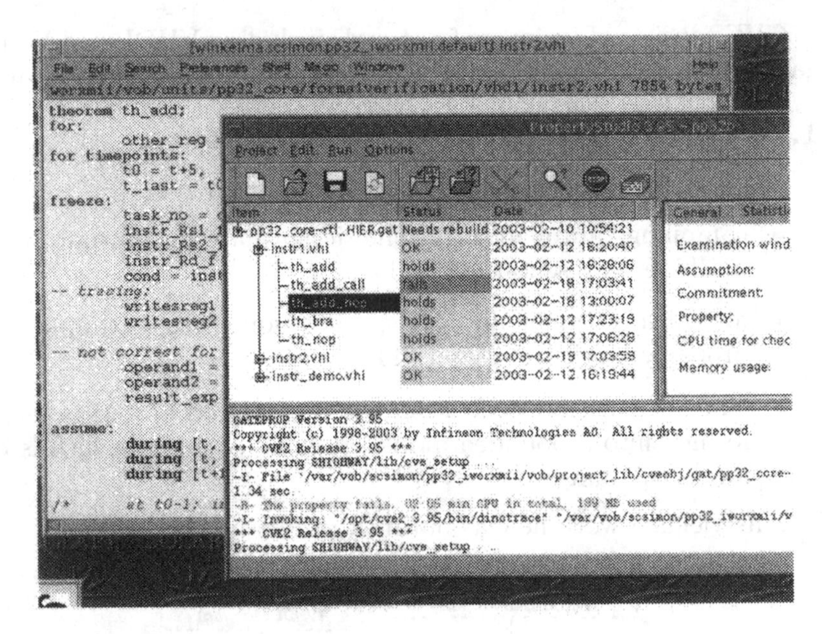

Figure 4.2. The CVE user interface for property checking

observation window of the appropriate length, in every run admitted by the design. The tool checks each property and, if it is found to be not universally valid, produces a counter-example. For blocks in the 30 to 100k gates range this check takes from seconds to a few minutes per property.

The basic concepts of the language are:

HDL flavour: The user chooses to write in either a VHDL or Verilog syntax, using familiar language constructs to quickly get up to speed with property writing.

Time steps: A property is written over a number of time steps, from time t (i.e. t+0) to a future time (e.g. time t+4 after 4 clock cycles). Consequently, there are a few time constructs in the language (e.g. at t, during[t,t+2], etc.).

Each property consists of a "prove" and an "assume" part.

Assume part: This part allows the designer to specify the working mode of the design under inspection. Assumptions such as no reset occurs at time t are typically necessary to investigate if the design exhibits a particular behavior. Further typical assumptions are at time t the input connection_request is high or there will be no write_request during time interval t+1 to t+5.

Prove part: This part of the property specifies expected behavior. Typical assertions in the prove-part are the grant output is set at time t+5 or the write_acknowledge output will somewhere be issued within time interval t+1 to t+3.

There are a number of language extensions that are designed to result in concise but intuitive properties, including data quantifiers, a powerful macro mechanism and time variables. As an example, here is a theorem for the ADD instruction of a processor.

```
theorem th_add;
freeze:
  Rs1           = natural'instr_Rs1(inst_i)      @ t0,
  Rs2           = natural'instr_Rs2(inst_i)      @ t0,
  Rd            = natural'instr_Rd(inst_i)       @ t0,
  cond          = natural'instr_cond(inst_i)     @ t0,
  task_no       = natural'"stack/task"           @ t0,
  operand1      = v_reg(task_no, Rs1)            @ t0,
  operand2      = v_reg(task_no, Rs2)            @ t0,
  /* compute expected result */
  result_exp    = (operand1 + operand2)          @ t0;
assume:
  general_assumptions;
  -- this is an ADD instruction:
  at t0:        instr_opcode(inst_i) = add_m;
  at t0 - 1:    satisfies(v_flags(task_no), cond);
prove:
  at t0 + 1:    v_reg(task_no,Rd)= result_exp(31 downto 0);
  at t0 + 1:    v_flags(task_no).zero
                  = to_bit(result_exp(31 downto 0) = 0);
  at t0 + 1:    v_flags(task_no).carry = result_exp(32);
end theorem;
```

Example ITL theorem specifying the ADD instruction of a processor.

2. Circuit Verification Environment: Underlying Techniques

2.1 From Property to Satisfiability

Here we outline the basic concept of transforming a property, together with the design, into a Boolean satisfiability problem. This is done in several steps.

First the design, given in a language such as Verilog or VHDL, is transformed into a finite state machine. The techniques for this are well known; however, large effort has been invested to actually develop an efficient and robust compiler covering the complete language subsets used in industry.

The finite state machine (FSM) thus obtained is now "unrolled", or "unfolded" to represent several successive time points, i.e. each next-state variable is fed into the state variable of a new copy of the FSM, until a sufficiently large time window is covered, depending on the timing constructs occuring in the property. Figure 4.3 illustrates the construction of this unrolled model.

A property describes a Boolean relation on this unrolled design, in terms of inputs, outputs and states at various time points. In Figure 4.4, a block representing such a relation is added.

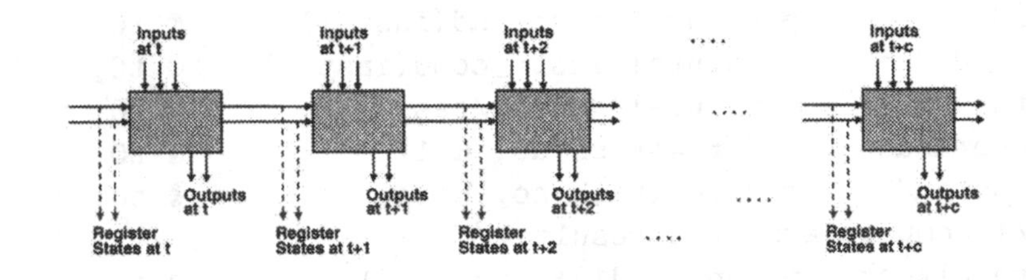

Figure 4.3. Unrolling

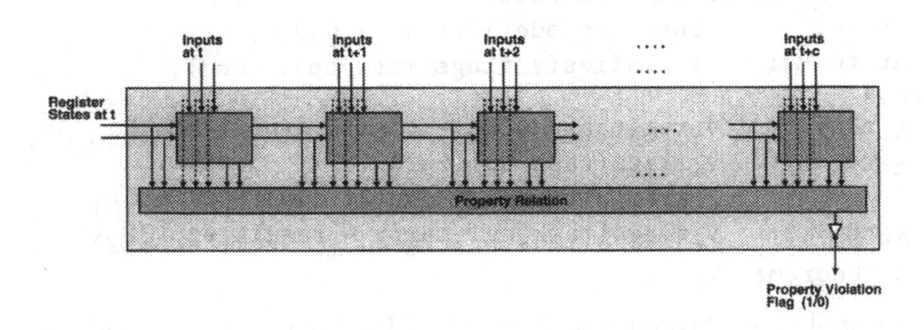

Figure 4.4. Bounded Model Checking Problem

Proprietary satisfiability algorithms are now imployed to decide the truth of the property.

If the proof fails (the problem is satisfiable), gateprop will provide a debug sequence in terms of the inputs and states required to demonstrate the failure - in case of the above example this would mean that the output will not be constant.

Debugging in gateprop is again via a waveform of inputs, states and outputs, as shown in Figure 4.5. However, gateprop does not use reachability analysis so it is possible that the design can start in an unreachable state at time t. It turns out that this theoretical draw-back is very rarely an obstacle in practical applications. If it does occur, the user normally adds a few assumptions to the theorem and re-runs the proof. Clearly, this may create new proof obligation to be discharged later-on, e.g. by proving a separate theorem for the remaining cases, or by proving an inductive invariant.

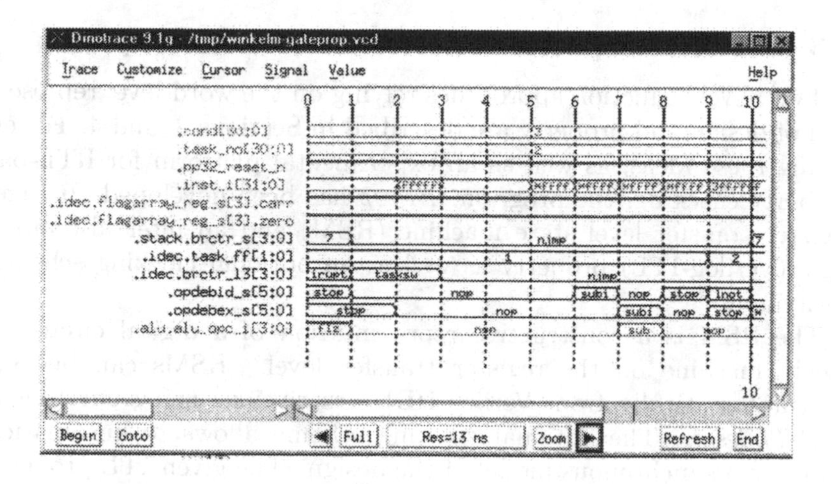

Figure 4.5. A wave-form showing a counter-example

2.2 Preserving Structure during Problem Construction

All formal approaches to verification, including property checking, use some kind of formal representation for a design. Structure, i.e. functional units and their operations, can only be exploited, if it is explicitly available in this formal representation. Hardware description languages, in which the functionality of a design is first formulated, operate on the register transfer level (RTL). On the register transfer level structure is explicit. It is prevalent in data types and functional units. Bitvectors (vectors of bits) are the prevailing data type. We exploit this structure to reduce the computational effort for solving property checking problems.

However, the standard gateprop flow destroys a lot of the structure, because the design is flattened into a bit-level representation. Hence the syntactic correlation of the single bits, available on the register transfer level, is lost. Therefore, the regular structure of a design is not explicit in its formal representation. To be able to exploit the structure of a design it must be preserved.

If, in contrast to the standard approach, the design is synthesized into a register transfer level representation, a word-level representation of the property checking problem can be constructed, and the structure is preserved. Now, structural information can be used to reduce the size of property checking problems before solving them. Such a reduction is of most practical use, if this reduction is fully automatic, and is seamlessly integrated into a tool.

2.3 The Experimental Platform RtProp

Two novel reduction approaches relying on the word-level representation of design and property are described in Sections 3, and 4. For evaluating these ideas, as well as an experimental platform for RTL-based property checking, the program *RtProp* has been developed. It takes a register transfer level state machine (RSM) and an interval-temporal-logic (Verilog-ITL) property as well as an optional clocking scheme as basic input.

The RSM is a constructive representation of a digital circuit as a Mealy machine on the register transfer level. RSMs can be generated automatically from Verilog-HDL descriptions using our tool *verilogRTL2rsm*. The optional clocking scheme allows to automatically compute a synchronous model of the design. The given ITL-property is also taken as input by *gateprop* which makes it easy to try RtProp on relevant industrial examples. The property is first translated into a basic form removing syntactic sugar, such that it is an expression over input output and state variables of the design annotated with time points.

The output and transition functions of the design as well as property are represented by bitvector terms [13, 24]. Bitvector terms are represented internally by fully collapsed term graphs, maximally sharing common subexpressions [26]. Normalization techniques such as constant propagation are built into the term generation process and are implicitly applied when terms are constructed.

From design and property a bitvector term with Boolean output is constructed which is constantly 1 iff the property holds in all states of the design.

Bitvector terms can be translated into vectors of Boolean terms as usual. Boolean terms can be checked for constantness using our composite prover, incorporating our state of the art proprietary BDD and SAT techniques. (For a similar approach see [25].) If counterexamples are generated they are automatically translated back to the RT-level and presented to the user in textual form.

Different reduction techniques are implemented and can be applied to the problem before solving it on the Boolean level. Two of the integrated reduction techniques, namely symmetry reduction and automated data path scaling, are described in Sections 3, and 4. Another idea implemented in RtProp can be found in [9].

The represented techniques as well as all steps described above (from reading the input files, through generating the problem, applying reduction and giving a counterexample if appropriate) are seamlessly integrated into the tool and require no user assistance.

3. Exploiting Symmetries

Analysis of the structure of property checking problems for regular designs such as memories, arbiters and bus systems, has shown that symmetry is an important structural feature, that can, and should be exploited to make the bounded model checking approach even more powerful.

3.1 Symmetry in Property Checking Problems

To solve the property checking problems generated by gateprop, as well as RtProp, it has to be checked whether some Boolean function is identical to 1 (for a similar approach see e.g. [5]). As described before this is mainly done using search procedures like SAT, which employ clever search heuristics. However, symmetries in search problems, and in particular in bounded model checking problems, make these search heuristics ineffective. This often leads to exponential behavior and thus to inefficiency, because symmetrical parts of the search space are considered several times. The idea of symmetry reduction is to prevent this.

Basically, there are two kinds of symmetry found in Boolean functions. The first is invariance under permutation of some variables (symmetrical variables), the second is invariance under assignment of different values to the same variable (symmetrical values of a variable).

The intuitive notion of symmetry is that of symmetrical variables. A Boolean function f is symmetrical in some variables x and y if f stays invariant under permutation of x and y. Consider for example the function f_1:

Example 4.1 (Symmetrical Variables)

$$\begin{aligned} g : \{0,1\} \times \{0,1\} &\rightarrow \{0,1\} \\ g(x,y) &\mapsto x \wedge y \end{aligned}$$

In g, x and y are interchangeable since $\wedge$ is a commutative operation. Hence x and y are symmetrical variables in g.

There are several published approaches on how to exploit this kind of symmetry, e.g. [14, 11, 21]. However, for our examples neither of them worked well since either such symmetries were not present, or if they were detected, the reduction did not show much effect.

In contrast, we will focus on symmetrical values, which are a generalization of equivalent values. We will first give a notion of equivalent values and their exploitation and then generalize to symmetrical values.

In general, if $f : D_1 \times \cdots \times D_n \to D$ is a function in n variables X with discrete domains and codomains, then the values a and b are *equivalent values* for a variable $x \in X$ iff the restrictions of f to $x = a$ and $x = b$ are the same, i.e. iff $f|_{x=a} = f|_{x=b}$. If all values for x are pairwise equivalent then the problem whether $f = 1$ is independent of x.

Let in the following $f : \{0,1\} \times \cdots \times \{0,1\} \to \{0,1\}$ be an n-ary Boolean function in n variables $X = \{x_1, \ldots, x_n\}$. The restrictions of f to $x = 0$ and $x = 1$ are the cofactors of f w.r.t. x. If the values 0 and 1 are equivalent for $x \in X$ (i.e. if $f_x = f_{\bar{x}}$ then $f = 1$ is independent of x. So, if we want to check whether $f = 1$ holds, it is sufficient to show either $f_x = 1$ or $f_{\bar{x}} = 1$. Conversely, if we know that $f_x \neq f_{\bar{x}}$ then $f = 1$ does not hold.

Now the notion of equivalent values is extended to symmetrical values. First consider the following example.

Example 4.2 (Symmetrical Values) As an example consider the function f_2 below.

$$\begin{aligned} h : \{0,1\} \times \{0,1\} \times \{0,1\} &\to \{0,1\} \\ h(x,y,z) &\mapsto ite(x, y \to (y \vee z), z \to (z \vee y)) \end{aligned}$$

Then $h|_{x=1} = y \to (y \vee z)$ and $h|_{x=0} = z \to (z \vee y)$. It is easy to see that both cofactors are equivalent, since they are both tautological (identical to 1). (However, in general (e.g. when h is a property checking problem) this is not so easy to see, i.e. it is in general a computationally complex task.)

Obviously, if a function is constant, renaming variables does not change this fact, since it is independent of all variables. For a Boolean function f and a permutation π of variables in f we have in particular $f = 1$ iff $\pi(f) = 1$. We can conclude that $f = 1$ holds iff $f|_{x=1} = 1$, and there exists some π fixing x such that $f|_{x=1} = \pi(f|_{x=0})$.

Example 4.3 (Symmetry Reduction) Consider h in the example above. By looking at the cofactors $h|_{x=1} = y \rightarrow (y \vee z)$ and $h|_{x=0} = z \rightarrow (z \vee y)$ it is evident that their representations are very similar. In fact the terms $y \rightarrow (y \vee z)$ and $z \rightarrow (z \vee y)$ have the same structure. They only differ in the names of variables. By permuting y and z they can be mapped onto each other. This means that π permuting y and z (while fixing x) we found such a variable renaming and can conclude: $h = 1$ iff $h|_{x=1} = 1$. Thus we can eliminate x from the problem and consider either cofactor further.

In the following the values 0 and 1 are called *symmetrical values* for x in f, iff the two cofactors of f w.r.t. x are *permutation equivalent* (symmetrical), i.e. if there exists a permutation π of the variables in f fixing x such that $f|_{x=1} = \pi(f|_{x=0})$. Note that if we can prove that there is no such π then f may not be constant and we can conclude that $f = 1$ does not hold. (Note that if π is restricted to be the identity the argument above is still valid, and we are back to the equivalent value case.)

Now let X be the set of all variables in f. Consider the situation where a set of k variables $X_k = \{x_1, \ldots, x_k\} \subset X$ is 'factored out' at the same time. Let $a_1, \ldots, a_k, b_1, \ldots, b_k \in \{0, 1\}$ then vectors $(a_1, \ldots, a_k)$ and $(b_1, \ldots, b_k)$ are called *equivalent value vectors* for the vector of variables $(x_1, \ldots, x_k)$ w.r.t. f, iff

$$f|_{x_1=a_1,\ldots,x_k=a_k} = f|_{x_1=b_1,\ldots,x_k=b_k}$$

(Here the functions $f|_{x_1=a_1,\ldots,x_k=a_k}$ and $f|_{x_1=b_1,\ldots,x_k=b_k}$ are k'th order cofactors of f w.r.t. X_k.) It is easy to see that equivalence of value vectors is an equivalence relation on the 2^k different value vectors, which are partitioned into equivalence classes.

Applying the same extension as in the case of one variable the vectors $(a_1, \ldots, a_k)$ and $(b_1, \ldots, b_k)$ are called *symmetrical value vectors* for the vector of variables $(x_1, \ldots, x_k)$ w.r.t. f, iff

$$\exists \pi \in Sym(X \backslash X_k) : f|_{x_1=a_1,\ldots,x_k=a_k} = \pi(f|_{x_1=b_1,\ldots,x_k=b_k})$$

(Note that π above can be extended trivially to a permutation of X, fixing the elements of X_k.) Symmetry of value vectors is an equivalence relation on the 2^k different value vectors, which are partitioned into equivalence classes. This equivalence relation is called the *symmetry relation* for f w.r.t. X_k.

Now, symmetrical value vectors for vectors of variables in some Boolean function, if known, can be used for preprocessing in property checkers. Let F_{X_k} be the set of all k-th order cofactors of f w.r.t. a set $X_k \subseteq X$ with $|X_k| = k$. Then $f = 1$ holds iff all cofactors in F_{X_k} are constantly 1, or, respectively, if they are all pairwise permutation equivalent and either of them is constantly one.

$$\begin{aligned} & f = 1 \\ \Leftrightarrow \quad & \forall f', g' \in F_{X_k} \exists \pi \in Sym(X \backslash X_k) : f' = \pi(g') \ \wedge \ \exists h' \in F_{X_k} : h' = 1 \end{aligned}$$

Example 4.4 (Exploiting Symmetrical Values) As an example consider a 256×16 memory and property describing its write behavior, as shown in Figures 4.6 and 4.7. The resulting problem is obviously symmetrical in all values for the bitvector of **ax** and **dx**, respectively. If this can be shown formally by a tool this knowledge can be exploited in such a way that only one 8×16-order cofactor of the respective Boolean function has to be considered. This reduces the search space by a factor of 2^{128}.

In general the symmetry relation is not a single equivalence class. Even if it is, sometimes just an approximation of the symmetry relation might be known. Then still only one cofactor for each equivalence class has to be checked. Let $P = \{P_1, \ldots, P_l\}$ be a partition of 2^k, and let $R = \{r_1, \ldots, r_l\}$ with $r_j = (a_{j1}, \ldots, a_{jk}) \in P_j, 1 \leq j \leq l$ be a set of representatives of each equivalence class. Then the following holds:

$$f = 1 \Leftrightarrow \bigwedge_{1 \leq j \leq l} f|_{x_1 = a_{j1}, \ldots, x_k = a_{jk}} = 1$$

Obviously the quality of the reduction depends on how good the approximation of the symmetry relation is and how easy it can be computed. Approaches for finding good approximate symmetry relations are described below.

3.2 Finding Symmetrical Value Vectors

The problem of finding symmetrical value vectors can be solved by showing permutation equivalence of the respective cofactors. In general, two n-ary Boolean functions f and g over variables $X = \{x_1, \ldots, x_n\}$ are *permutation equivalent* (symmetrical) iff there is a permutation $\pi \in Sym(X)$ of the variables X such that $f = \pi(g)$, i.e. $f(x_1, \ldots, x_n) = g(\pi(x_1), \ldots, \pi(x_n))$ for all $x_1, \ldots, x_n \in \{0, 1\}$. Such a permutation is called a *variable renaming.* The problem of finding π such that for two functions f and g, we have $f = \pi(g)$ is known as the permutation equivalence problem. For Boolean functions it is also known as Boolean isomorphism problem.

```
'define MEM_WIDTH 16
'define MEM_DEPTH 8
'define MEM_CELLS 256

module memory (cs, rw, rs, a, di, do, clk);
   input cs;
   input rw;
   input rs;
   input ['MEM_DEPTH-1:0] a;
   input ['MEM_WIDTH-1:0] di;
   input                 clk;
   output ['MEM_WIDTH-1:0] do;
   reg ['MEM_WIDTH-1:0]    m ['MEM_CELLS-1:0];
   reg ['MEM_WIDTH-1:0]    o;

   integer                 i;

   always @(posedge clk)
      begin
         if (rs)
            begin
               for (i = 0; i<='MEM_CELLS-1 ; i = i+1)
                  begin
                     m[i]='MEM_WIDTH'd0;
                  end
               o = 'MEM_WIDTH'd0;
            end
         else   // if (clk)
           if (cs)
              begin
                 if (rw)
                    m[a]=di;
                 else
                    o = m[a];
              end
           else
              if (! rw)
                 o = 'MEM_WIDTH'd0;
      end // always @ (posedge clk or negedge reset)
      assign do = o;
endmodule // memory
```

Figure 4.6. Verilog-HDL Code of Memory

```
theorem write;

for:
        ax = 0..MEM_CELLS-1,
        dx = 0..MEM_WIDTH-1;
freeze:
        dixt = di[dx]@t;
assume:
        at t: rs == 1'b0;
        at t: cs == 1'b1;
        at t: rw == 1'b1;
        at t:  a == ax;
prove:
        at t+1:  m[ax][dx] == dixt;
end theorem;
```

Figure 4.7. Write Property for Memory in Verilog-ITL

Example 4.5 (Symmetrical Functions) As an example consider the functions f and g:

$$\begin{aligned} f : \{0,1\} \times \{0,1\} &\rightarrow \{0,1\} \\ f(x,y) &\mapsto (x \wedge \neg y) \vee (\neg x \wedge \neg y) \\ g : \{0,1\} \times \{0,1\} &\rightarrow \{0,1\} \\ g(x,y) &\mapsto (y \wedge \neg x) \vee (\neg y \wedge \neg x) \end{aligned}$$

They are identical after renaming x with y and y with x in either function, i.e. for $\pi = (x \mapsto y, y \mapsto x)$ we have $f(x,y) = g(\pi(x), \pi(y))$. Note that the functions f' and g' below are also symmetrical by the same π.

$$\begin{aligned} f' : \{0,1\} \times \{0,1\} &\rightarrow \{0,1\} \\ f'(x,y) &\mapsto \neg y \\ g' : \{0,1\} \times \{0,1\} &\rightarrow \{0,1\} \\ g'(x,y) &\mapsto \neg x \end{aligned}$$

The functions f and f' are actually the same. Therefore they are also symmetrical. (The same holds for g and g'.)

It is well known that the permutation equivalence problem for Boolean functions is Co-NP-complete [1]. So, the question is why should it be easier to solve the permutation equivalence problem for cofactors than proving their constantness individually.

The answer is twofold. On one hand it is to be expected that, for Boolean functions generated when property checking, self similarities, prevalent in the structure of design and property, are reflected in the representation of this function (as in h). By choosing the set X_k appropriately this should lead to similar representations for the cofactors.

On the other hand we do not really need to decide the permutation equivalence of cofactors. Instead it is often sufficient to employ semi-decision procedures. Using a semi-decision procedure the set of all possible value vectors, and hence the set of all cofactors F_{X_k} w.r.t. X_k is partitioned into equivalence classes of an approximate symmetry relation. In this case we need to prove $f' = 1$ for only one representative f' of each equivalence class.

Such procedures are for example (semi-) decision procedures for Boolean equivalence (where π is the identity), as known from combinational equivalence checking or certain rewrite heuristics. Structural symmetry of cofactors is another criterion implying permutation equivalence. Obviously different approaches can be combined in order to improve a given symmetry relation. It is also possible to compute symmetry relations for two disjoint sets of variables X_k and X_l separately and to join them yielding a symmetry relation for $X_k \bigcup X_l$.

Since many of these approaches benefit from structural similarities in the function representation our implementation lifts the idea to the register-transfer-level, i.e. instead of a Boolean function we consider a bitvector function f' with Boolean codomain. The grouping of Boolean variables to bitvector variables and the representation of f' is naturally derived from the bitvector variables of the RTL representation of design and property. This ensures that structural self similarities are preserved as much as possible by representing the Boolean function as bitvector term (comparable to an HDL expression). It also gives a guideline on how to choose X_k which is done along the bits of bitvector variables.

For computing approximate symmetry relations we developed new methods operating on sets of bitvector functions. They combine different rewrite heuristics for normalization and reduction of complex bitvector term graphs with isomorphism and automorphism procedures for specially labeled directed acyclic graphs. These graphs originate in the term graph representation of the bitvector functions (cofactors) in question. If two graphs for two functions are isomorphic, then the respective term graphs have the same structure modulo permutation of arguments of commutative function symbols. The automorphism procedure allows to check many graphs (and thus many functions) for isomorphism at the same time. A detailed description of the approach is out of our scope here, but some of the ideas are presented in more detail in [10].

3.3 Practical Results

The reduction approach described above has been integrated into the tool RtProp and has been used on industrial designs, in many cases showing considerable performance gains.

To make the approach work practically, different rewrite heuristics for bit-vector terms have been developed. Furthermore different degrees of structural symmetry can be discovered by specially adopted graph algorithms working on bitvector terms. Interleaving them leads to fast and reliable preprocessing techniques for property checking, exploiting symmetrical values.

For normalization a prototypical rewrite system was implemented. As graph automorphism engine an implementation of [22], optimized for sparse graphs, was integrated. Currently the following reductions are implemented.

1 Rewrite heuristics to be applied on top of the built-in normalization can be specified in form of rule specifications contained in an additional file. Two different rewrite strategies (top-down and bottom-up rewriting) can be selected and are combined. Rule specifications can be selected for application in either or both strategies. Rewriting is implemented as term graph rewriting on fully collapsed term graphs such that common subexpressions are only treated once. During top-down (bottom-up) rewriting subterms are normalized in reverse topological (topological) order until all subterms are in normal form.

2 Symmetry reduction can be performed for the bitvector variables of the problem belonging to 'for'-variables in the given ITL-property. Different strategies can be selected by options, for example whether rewriting should be applied to the whole bitvector term before generating cofactor terms or not. During symmetry reduction equivalence classes for the values of these selected variables are subsequently improved. First simple heuristics are tried, if no more reduction is achieved more sophisticated and more expensive methods are applied.

One of our industrial applications is a 256x55-bit 2 way set associative Tag-RAM as they are commonly used in memory caches. It consists of about 500 lines of Verilog source code and could not be scaled down by hand easily. Therefore it was desirable to verify the design as is. It contains a lot of control logic to organize the data transfers and some arithmetic for determining whether a hit of miss occurred. It was verified that the read and write operations work correctly and the content of the tag memory does not change if the memory is not enabled. The properties are parameterizable in the number of memory cells, between 2 and 256. The properties are specified over 8 time steps. Up to 6 variables have been selected for elimination. The advantages of the reduction approach are tremendous for all three properties.

The first property ensures that the memory content does not change unwanted. For the configurations proving this property for 2 to 32 addresses (address input values) and 2 to 32 memory cells all cofactors were collapsed to a single one which was proven almost in no time. In the other cases not all cofactors were identified leaving 64, 128 and 256 small problems. The reduction time increased with the number of cofactors, except the transition from 128 to 256. Here the reduction procedure used a different strategy due to the large number of cofactors (It gave up earlier on some reductions). Proving the property without reduction required more than three hours CPU-time. Using reduction this time was reduced to about 7 minutes.

The situation is comparable for the other properties. Using reduction the last property has been proven for the first time (however, spending 14 hours CPU time). It could not be proven before using *gateprop*. Here, the remaining cofactors were often symmetrical for different configurations since just the property was scaled up.

Note that proving many of the reduced properties was intractable for BIMC at the time our research efforts on symmetry reduction started. Only due to improvements on the SAT-solver implementation using newest results [23] the ratio changed somewhat in favor of not reducing the problem before solving it. However, this effect was not strong enough. Still the reduction approach leads to a speedup by one to two orders of magnitude in relevant cases, proving a property intractable for the standard approach even now.

At first sight, our symmetry reduction scheme has some similiarities with the one proposed in [15]. There so called scalar set variables are used to indicate possible symmetries in the state space of a design. However these scalar set variables can be used in only a few syntactic contexts in order to maintain the structural symmetry of the design. Therefore the approach can not be used within the context of Verilog- and VHDL-HDL verification. In contrast, our approach has many advantages. Since it is defined semantically it does not suffer from syntactic restrictions on the variables containing symmetries as in [15]. Although the 'for' variables of the property are taken as candidates for symmetry reduction, this is not necessary. Any possible variable in the function to be checked can be taken. In fact, symmetry relations allow even much more complex relationships, for example between values of different variables. At the same time, syntactic self similiarities in the design and property can be exploited as efficiently.

4. Automated Data Path Scaling to Speed Up Property Checking

In the following, another formal high-level technique is described which is used in CVE in order to speed up property checking runtimes. The core functionality of this approach is based on a simple and straightforward idea. High-level design specifications of digital circuits contain the structural information on how single bits are arranged to represent word-level signals and which individual bits belong to the same word-level signal. The information about the widths of data path signals and about word-level data flow is available and can be exploited.

Under specific conditions, it is possible to replace an n-bit data path of a circuit design by an m-bit data path, with $m < n$, and then to use the scaled and smaller version of the design for verification instead of the original one without altering verification results. Such data path scaling is a classical means for attacking the state space explosion problem.

Reduction of data path widths is typically tried if verification of an ASIC which includes an n-bit data path takes too long or cannot be completed due to reasons of computational complexity. So far, data path scaling was often done *manually* based on experience and intuition of the circuit designer, usually without having the (formal) guarantee that the property which was considered and had to be verified really was independent of the width of the data path and that the chosen amount of scaling did not falsify verification results. Moreover, such manual modifications usually required intensive rewriting of the HDL code as shrinking the width of a data path causes additional side-effects. If, for example, a 32-bit bus is replaced by a bus of smaller width, say 16-bit, then the width of each signal which accesses the bus by read or write operations must be scaled too. Side-effects can go even farther if such a signal is the concatenation of several other smaller signals. Consider a design where a 24-bit signal reads information from the 24 most-significant bits of the bus while another 8-bit signal reads the 8 least-significant bits. At the outset, it is not clear how reducing the bus-width to 16 bits affects the two signals which read from the bus and how scaling has to be applied to them.

The approach implemented in CVE was presented in [20] and allows for a fully automated scaling of data path widths. The technique efficiently analyzes word-level data flow in RTL design descriptions with respect to a specified property. Designs are automatically scaled down by reducing signal widths before property checking, while guaranteeing that the property holds for the scaled model if and only if it holds for the original design. The reduced model of the circuit is used as input instead of the original design, thus speeding up property checking runtimes and allowing larger design sizes to be verified.

4.1 Bitvector Satisfiability Problems

The data path scaling technique is based on data flow specification by means of *formal bitvector theories* (see e.g. [12, 24]). Bitvector theories have proven to be an adequate means of describing digital hardware and related Bounded Model Checking problems at a higher level of design abstraction. Bitvectors are array-like structures of finite width over a two-valued domain, which can be used to model multi-bit circuit signals. Word-level data flow and control flow aspects of digital designs can be characterized by bitvector equations in a way, such that design properties can be verified by determining satisfiability of such equations. Several decision procedures have been investigated which determine satisfiability or validity of bitvector equations, see e.g. [3, 4, 6, 12, 13, 24, 18]. However, either the expressiveness of the term languages and the bitvector theories which are used is rather limited, or the performance of the decision procedures cannot compete with SAT and BDD based property checking when applied to large real world circuit designs.

Instead of trying to directly solve bitvector equations, the approach implemented in CVE utilizes the high-level information contained in the bitvector terms in order to compute a second corresponding and equivalent system of bitvector equations for which then satisfiability is determined by using conventional SAT and BDD based methods. Thus, a high-level abstraction technique for systems of bitvector equations is established, which is characterized in the way that the SAT problem which is related to the second system is smaller than the SAT problem related to the original system and therefore generally can be decided much faster and with less computational effort (see Figure 4.8 for an illustration of the basic concept).

The mathematical framework of this abstraction technique is the formal satisfiability problem **BvSAT** for bitvector functions and bitvector disequalities, which was first presented in [19] and which is a generalization of the SAT problem from Boolean variables to bitvectors of finite width. Satisfiability of systems of bitvector equations can be reduced to satisfiability of instances of **BvSAT**.

The data path scaling is based on a size reduction of **BvSAT** problems by means of a formal one-to-one correspondence between given instances

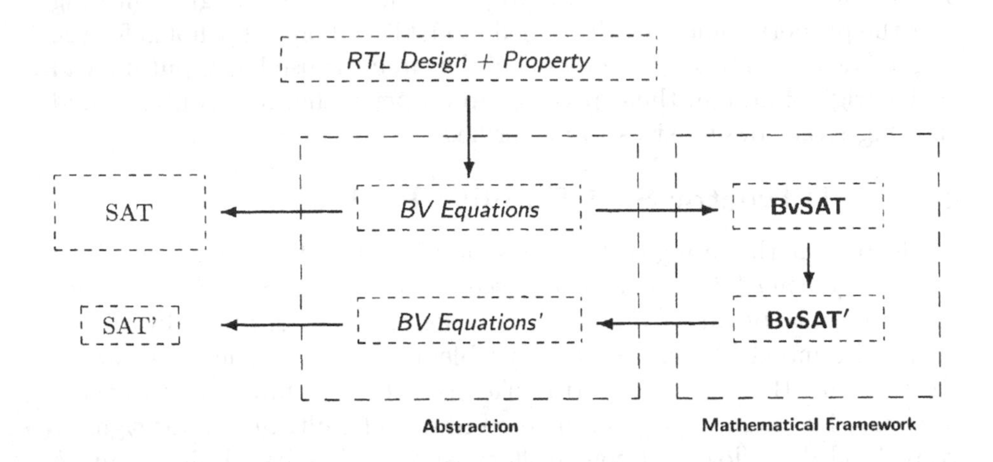

Figure 4.8. Basic Concept

of **BvSAT** and related instances over bitvector domains of *smaller* width. This correspondence maps satisfiable instances onto smaller satisfiable instances, and unsatisfiable ones onto smaller unsatisfiable instances (i.e. the correspondence preserves satisfiability in a one-to-one fashion). It is used in order to reduce a system of equations over bitvectors of certain widths into an equivalent system over bitvectors of smaller widths, while preserving satisfiability of the equations in a one-to-one fashion. The reduction is based on symbolic analysis of word-level data flow and detection of uniform data dependencies. The reduction technique furthermore provides an efficient way to compute satisfying solutions of the original system from satisfying solutions of the reduced system.

Satisfiability of bitvector equations can be checked in the Boolean domain by transforming systems of bitvector equations into Boolean formulae, i.e. into instances of propositional SAT, and afterwards applying bit-level satisfiability checks, like SAT and BDD based procedures. Thus, bitvector formalisms are ideally suited for combining BMC and high-level verification techniques. The complexity of determining satisfiability of Boolean formulae generally depends on the number of Boolean variables occurring in the formulae. When systems of bitvector equations are transformed into SAT problems, Boolean variables are generated for each bit of each bitvector variable. Thus, the complexity of checking satisfiability of systems of bitvector equations directly depends on the overall number of bits of all input, internal and output signals occurring in a design, i.e. on the sum of the widths of all bitvectors occurring in the equations. As a consequence, width reductions can have a significant impact on the runtimes of satisfiability checkers.

4.2 Formal Abstraction Techniques

In general, abstraction techniques implement the following approach. Instead of directly solving a given verification problem P, a smaller or simpler instance $P' := \mathcal{T}(P)$ is computed in which information that is not relevant for solving the verification problem is abstracted and which is then solved by conventional methods.

Depending on the degree of reduction or simplification between P and P', solving P' can possibly be done faster and might require significantly less resources. It has to be ensured that computing P' from P preserves certain criteria as far as solvability is concerned. In this context, an abstraction technique $\mathcal{T}$ is said to be *one-to-one* if, for all problem instances P, solvability of $P' := \mathcal{T}(P)$ is related to solvability of P in a one-to-one fashion, i.e. if the abstract problem is solvable *if and only if* the original problem is solvable. Since we consider satisfiability problems, we have:

$$\mathcal{T} \text{ is a one-to-one abstraction:} \quad (P' \text{ satisfiable} \iff P \text{ satisfiable})$$

If the domains of abstract and original problem differ, then abstractions usually provide an additional transformation τ which computes solutions of the original problem from solutions found on the abstract problem instance, i.e. if s is a satisfying solution of $P' := \mathcal{T}(P)$, then $\tau(s)$ yields a satisfying solution of P.

Thus, solving the original problem can completely and efficiently be replaced by solving the abstract problem instance, provided that the total amount of time for computing the abstract instance and then solving it is still faster than solving the original problem. For example, conversion of Boolean SAT formulae from CNF to DNF yields a one-to-one abstraction with respect to satisfiability. As far as complexity is concerned, deciding satisfiability of CNF formulae is NP-complete, whereas satisfiability of DNF formulae can be determined in polynomial time. However, whether there exists an efficient computation of the abstraction itself, is still an open problem which is equivalent to the P=NP problem.

If an abstraction is not one-to-one, then for each solution s found for P', an additional consistency check has to be performed which inspects if $\tau(s)$ indeed is a solution of P or not (a so-called *false-negative*). Such an abstraction still might be of interest if establishing a one-to-one abstraction is not possible, but finding solutions s of P' and performing the consistency check for $\tau(s)$ is fast. In such a case, abstraction is usually combined with guided-search techniques on the solution space of P'. For each solution that is found, a consistency check is performed, and the search is continued if this check fails. Yet, the amount of reduction of the problem size achieved by such an abstraction must justify the additional costs for validating solutions found for the abstract problem instance.

One-to-one abstractions are highly attractive in digital hardware verification because reduced or simplified problem instances can significantly increase the performance of existing verification tools. If the abstract problem instance is specified in the same formalism which is used for the original problem, then one-to-one abstractions can easily be embedded in existing verification flows without having to modify the underlying verification techniques. Additionally, abstractions which operate on RT-level can incorporate and utilize all high-level information which is available in the problem specification.

4.3 Speeding Up Hardware Verification by One-To-One Abstraction

The method used in CVE implements a fully automated word-level abstraction technique, which operates as a preprocess for property checking of digital hardware. A one-to-one RTL abstraction of a digital design is computed in which the widths of word-level signals are reduced with respect to a property.

Given an RTL specification of a digital design and a property φ, a reduced RTL model is generated in which each word-level signal $\boldsymbol{x}$ is replaced by a corresponding shrunken signal of width $m_x \leq n$, where n denotes the original width of $\boldsymbol{x}$. The method establishes a one-to-one abstraction, i.e. φ holds for the original design if and only if φ holds for the reduced model. False-negatives cannot occur.

Design and abstract model differ from each other only as far as signal widths are concerned. Each m_x is the minimum number of bits for $\boldsymbol{x}$ which is necessary and sufficient in order to establish a one-to-one abstraction for φ and a reduced model of the abovementioned type. The width of each signal in the abstract RTL model is minimal with respect to the design, the property φ, and the abstraction technique we propose (i.e. by solely changing signal widths).

Furthermore, a post-processing of counterexamples is provided. If φ does not hold for the abstract RTL, and if a counterexample in terms of value assignments to its input signals is found, then a counterexample for the original circuit can be computed. The verification task itself is completely carried out on the scaled-down version of the design. If the property fails, then CVE computes counterexamples for the original design from counterexamples found on the reduced model.

The method also strictly separates the pre-/postprocessing of design and counterexample and the property checking process itself. Thus, the approach itself is independent of the concrete realization of the property checker and can be combined with the whole variety of property checking techniques implemented in CVE.

Moreover, if preprocessing yields that no reduction is possible for a given design and a property, then abstract model and original design are identical. Thus, the verification task itself is not impaired by using the proposed abstraction as a preprocess.

Linear signal width reductions result in exponentially smaller state spaces. A linear reduction of a signal's width from n bits down to m bits, $m < n$, causes an exponential reduction of the size of the induced state space from 2^n down to 2^m. Hence, the proposed abstraction to a great extent can have influence on the verification runtimes and can significantly speed up property checking. Furthermore, state space reductions allow larger design sizes to be verified. Experimental results on large industrial circuits have demonstrated the applicability and efficiency of the proposed method.

4.4 Data Path Scaling of Circuit Designs

Data path scaling is used as a preprocessing step in the verification flow. The technique implemented in CVE exploits structural high-level design information and absorbs the informational gap between RTL and bit-level by establishing a fully automated scaling of the design.

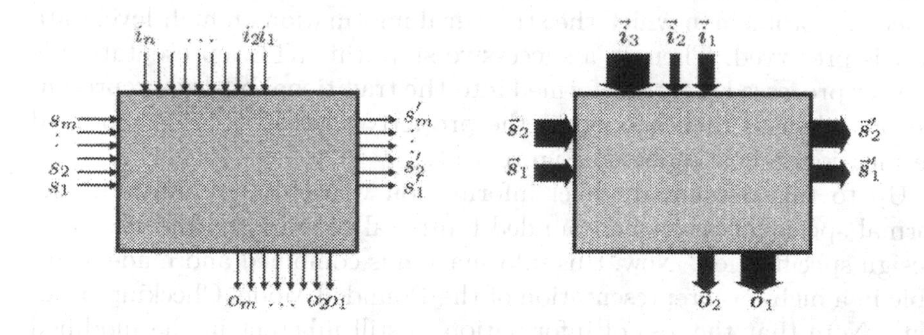

Figure 4.9. Bit-Level Design Specification vs. Word-Level Design Specification

Figure 4.9 demonstrates one of the differences between high-level and bit-level specifications of digital designs. The information which collections of single bits resemble high-level circuit signals is present in RTL specifications, but is lost on bit-level. As a consequence, the specification of data flow on bit-level lacks this information, too. Data flow can only be specified in terms of single-bit signals and Boolean logic gates, whereas on RT-level the data paths of a design are characterized by multi-bit busses and high-level operators and modules, as indicated in Figure 4.10.

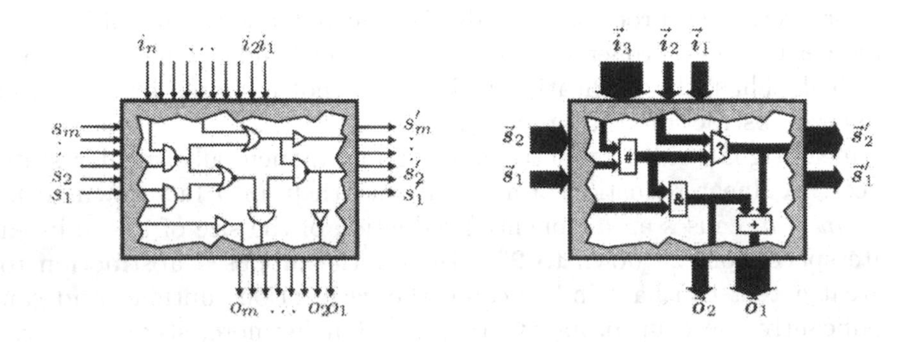

Figure 4.10. Bit-Level Data Flow vs. Word-Level Data Flow

The implemented abstraction technique exploits high-level information on multi-bit signals and high-level information on multi-bit data flow. In order to be able to do so, a prerequisite is to make this type of information available in an intermediate preprocessing stage of the verification flow. Therefore, in a first conceptual step, the conventional frontend is replaced by a new frontend which, instead of a bit-level representation, generates an RTL representation of the Bounded Model Checking problem in which the structural information on high-level data flow is preserved. Then, in a successive step, this RTL representation is further processed and transformed into the traditional bit-level representation, which is then handed to the property checker. The so modified verification flow is shown in Figure 4.11.

Up to this point, high-level information was only contained in the formal specification of the bounded temporal property and in the HDL design specification. Now, this information is combined and made available in a high-level representation of the Bounded Model Checking problem. Note that the loss of information is still inherent in the modified verification flow. It is deferred and now occurs in the newly introduced transformation step.

In this context, Bounded Model Checking problems are represented on RT-level by formal systems of bitvector equations. For a given design D and a formal property φ the new frontend synthesizes a system

E of bitvector equations such that the corresponding Bounded Model Checking problem $\hat{\varphi}$ is satisfiable if and only if E is satisfiable, i.e.

$$E \text{ is satisfiable} \iff \text{property } \varphi \text{ does not hold for design } D \quad (4.1)$$

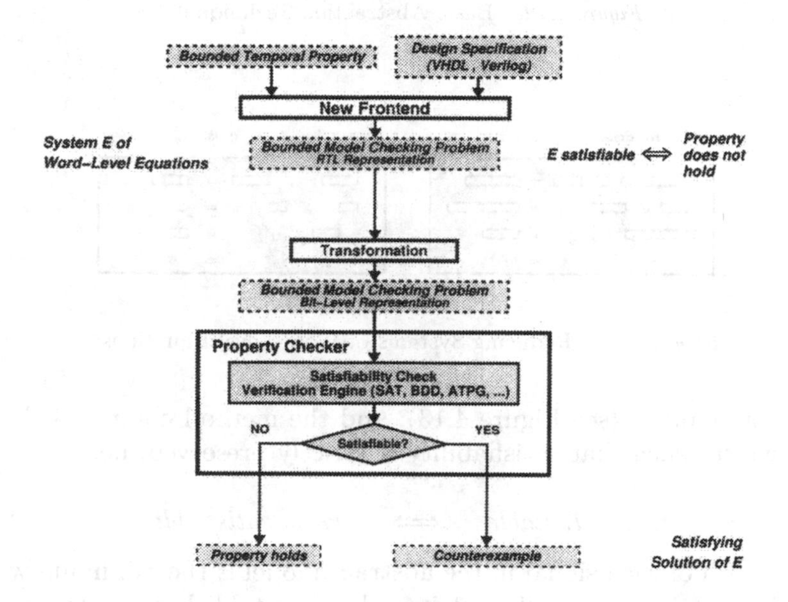

Figure 4.11. Modified Property Checking Flow

Bitvector variables of E correspond to multi-bit circuit signals of D. Each signal and each bitvector variable $\boldsymbol{x}$ has a fixed width $n \in \mathbb{N}_+$ (which in the following is sometimes annotated as a subscript in square brackets) and takes bitvectors of respective length as values. The later transformation of E into a bit-level representation of the Bounded Model Checking problem generates one bit-level variable for each bit of a bitvector variable. Thus satisfying solutions of E directly correspond to satisfying solutions of $\hat{\varphi}$ and vice versa, and yield counterexamples for φ and D (see Figure 4.11). The abstraction technique establishes a preprocessing step. The system E of bitvector equations is taken and analyzed, and a second system E' is computed which is then used for property checking instead of the original system E. The system E' is generated by replacing each word-level signal $\boldsymbol{x}$ of E by a corresponding shrunken signal of width $m \leq n$ (where n denotes the original width of $\boldsymbol{x}$).

The original system E and the abstract model E' differ from each other only as far as signal widths are concerned. All other data flow aspects, like for example operators or term structure,

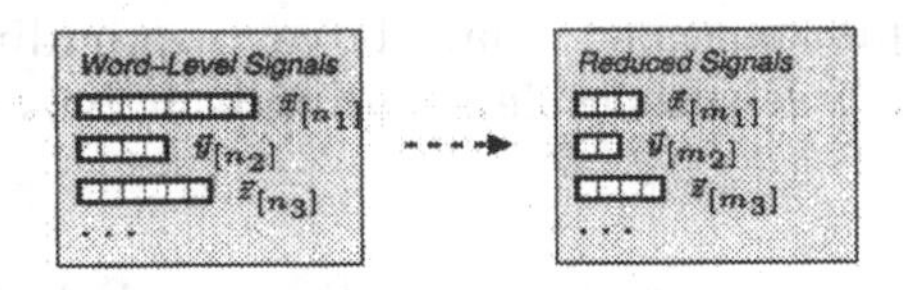

Figure 4.12. Basic Abstraction Technique

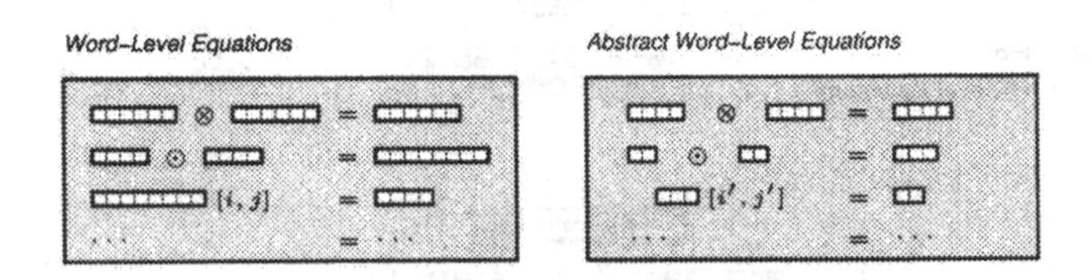

Figure 4.13. Reducing Systems of Word-Level Equations

are not changed (see Figure 4.13), and the method computes the reduced widths such that satisfiability is strictly preserved, i.e.

$$E \textit{ is satisfiable} \iff E' \textit{ is satisfiable} \tag{4.2}$$

The width of each signal in the abstract model is the minimum width which is necessary and sufficient in order to establish a one-to-one abstraction with respect to design, property, and the considered reduction technique (E' differing from E only by reduced variable widths). The reduced system E' then corresponds to a scaled version of the original design D, and scaling is understood in terms of strictly preserving the general data flow of D except for a reduction of the widths of the data paths, as illustrated in Figure 4.14.

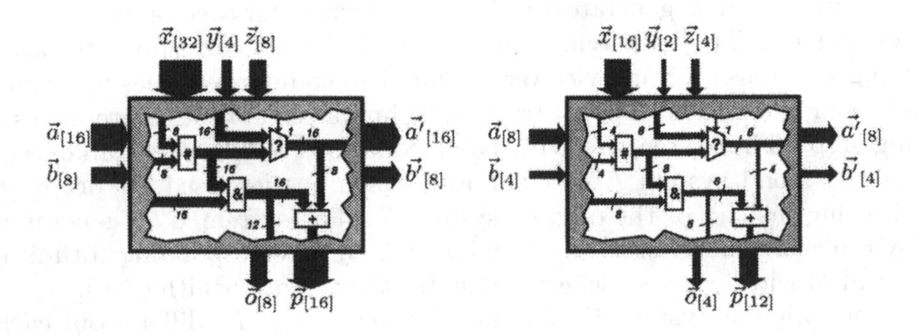

Figure 4.14. Original and Scaled Design

The amount of reduction that can be achieved is determined with respect to the circuit property that is to be verified. The scaled version of the design is then used for verification instead of the original one, and (4.2) yields:

$$\begin{gathered}\text{the property holds for the original design}\\ \Longleftrightarrow\\ \text{the property holds for the scaled design}\end{gathered}$$

If the property does not hold, then, considering the modified verification flow, the counterexample which is returned by the property checker is a counterexample for the scaled version of the design. The abstraction technique which is presented in this thesis adds an additional postprocessing step to the verification flow. In this step, the reduced counterexample is taken and a counterexample for the original design is generated. The proposed reduction technique provides an easy generation of satisfying solutions of E from satisfying solutions of E'. Figure 4.15 illustrates how the proposed abstraction technique is integrated in the verification flow.

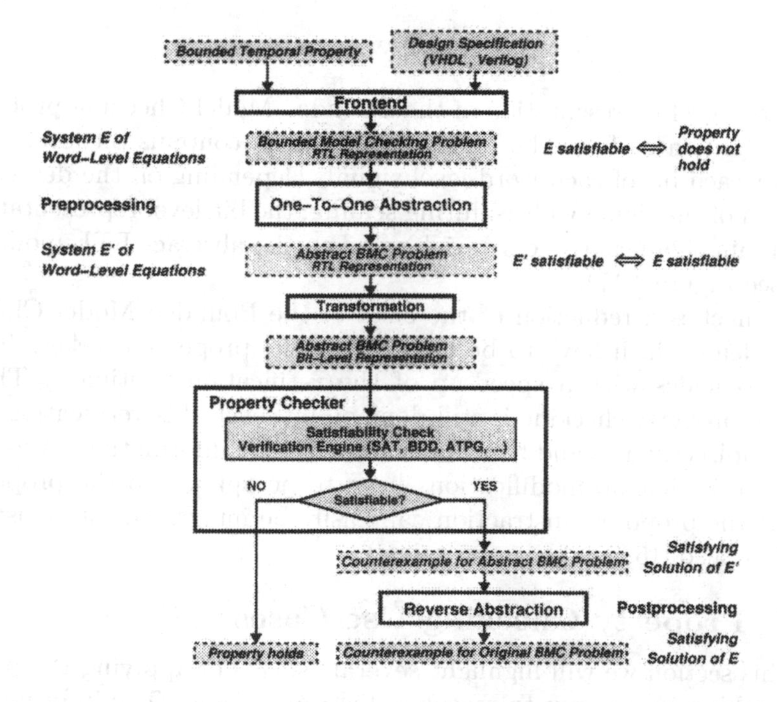

Figure 4.15. Property Checking Flow with High-Level Abstraction

The process of scaling down signal widths is separated into two subsequent phases. First, the coarsest *granularity* of each word-level signal is computed as determined by the structural data dependencies of E. A granularity is a separation of a signal into several contiguous parts which indicate the coarsest possible subsumptions of individual bits of the signal which are processed on the same data path.

Then, for each such part, the necessary minimum width is computed which guarantees that satisfiability of E and E' correspond to each other in a one-to-one fashion. According to these computed minimum widths, the reduced width for the corresponding signal is reassembled. The basic concept of this technique is shown in Figure 4.16.

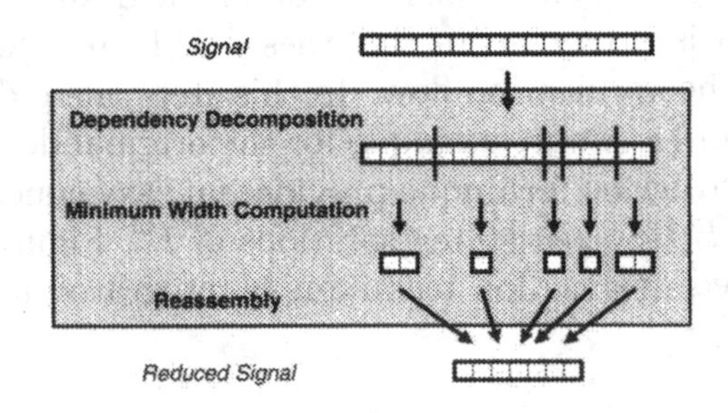

Figure 4.16. Signal Width Reduction

The bit-level representation of the Bounded Model Checking problem which is generated from the RTL representation contains bit-level variables for each bit of each word-level signal. Depending on the degree of reduction of the signal widths during scaling, the bit-level representation can contain significantly less variables when the abstract RTL model is used (see Figure 4.17).

The effect is a reduction of the sizes of the Bounded Model Checking problems which have to be handled by the property checker. This aspect coincides with a speed-up of the verification runtimes. Thus, although property checking is still done on bit-level, this verification approach indirectly uses and benefits from high-level information. Another advantage is that no modifications have to be applied to the property checker; the proposed abstraction can easily be integrated into existing flows (see also [16, 17] for an overview).

5. Property Checking Use Cases

In this section we will highlight several aspects of applying the property checking technology to recent industrial designs. The information given is based on more than 20 application projects carried out over the recent few years both at Infineon and Siemens. In most cases, the

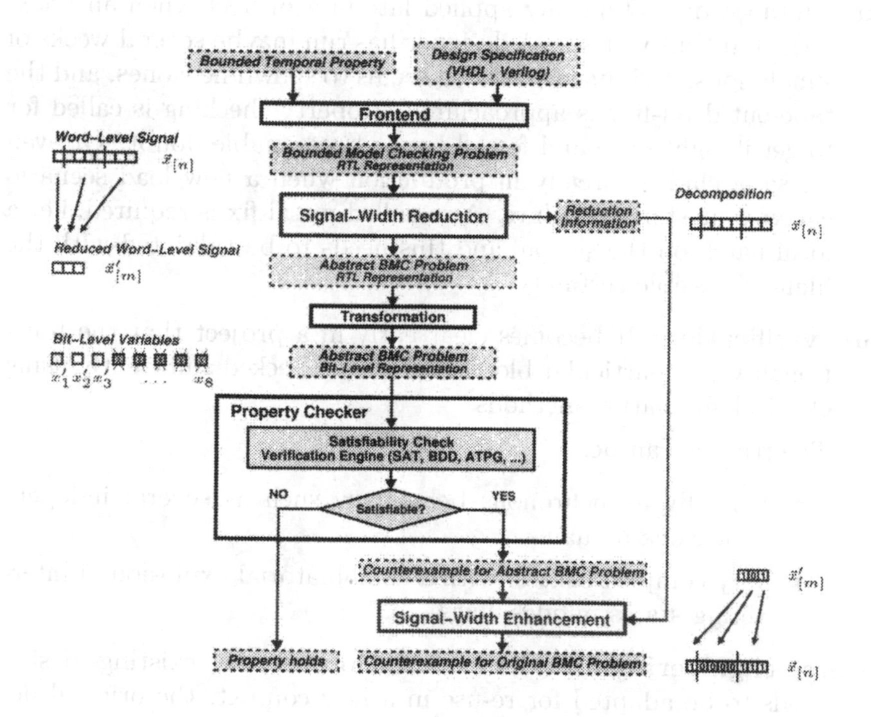

Figure 4.17. Speeding Up Property Checking by Automated Data Path Scaling

verification effort has initially been driven by an experienced verification engineer from the CVE team. After he has produced a first set of properties and detected the first bugs, typically the designers gradually adopt the method and start writing properties by themselves.

The size of the applications carried out, measured in verification effort, ranges from a few days up to several person-years. The largest project consists of 40 blocks verified with 660 properties.

Many different *use-cases* have been identified, differing in the particular project situation and the goals of the verification:

Hard debugging is typically applied late in a project when an unexpected difficulty arises. A designer has run maybe several weeks of simulations, each uncovered bug seems to spawn new ones, and the tape-out dead-line is approaching. Property checking is called for to get it right once and for all beyond reasonable doubt. Or even worse, a chip is already in production when a new load scenario causes a mysterious failure. A so-called metal-fix is required, i.e. a local patch on the silicon, and this needs to be validated with the highest possible certainty.

Hard verification: It becomes clear early in a project that the functionality of a particular block can not be checked sufficiently using classical simulation methods.

The reasons can be:

- partially asynchronous behaviour, such as several independent clock domains,
- very complex control with a combinatorial explosion of interacting states, modes, flags etc.

Reverse engineering and documentation: An existing design needs to be adapted for re-use in a new context, the original designer is not available, and the documentation turns out to be less detailed than required. A property suite is needed to unequivocally document the function of the block, its operation conditions as well as its communication protocols.

Early design validation is in a way the most mature application scenario. When starting a new ASIC design project, a deliberate decision is taken to spend a considerable portion of the validation budget for formal block-level verification. For each block either simulation or property checking is used as verification approach. The investment is paid off by the high quality of the blocks, allowing a smooth system integration.

For each of these scenarios we have experienced several instances. We will discuss examples for reverse Engineering and documentation as well as early design validation in sections 4.1 and 4.2, respectively. Section 4.3 will summarize the overall experience in terms of measurements for productivity.

The approach also provides for the correct integration of blocks and for the proper interaction of a few blocks. In addition, the property suite constitutes a precise, compact and intuitive documentation of the block under development.

5.1 Application Example: Reverse Engineering

This section describes the application of the gateprop property checker to a reverse engineering problem for a communication design used to convert AAL2 packets into ATM frames. (AAL2 is one particular communication protocol in the ASM Asynchronous Transfer Mode standard, see [2]).

5.1.1 Functionality. The design receives data packets (SDUs, service data units) from an array of DSPs and sends them to a dedicated ATM channel. While incoming packets have varying length, the resulting ATM packets (PDUs, protocol data units) have constant length of 48 bytes, such that several cases of combining and splitting need to be dealt with. Also one extra input channel for test purposes as well as one extra output channel for a buffer loop are taken into account.

The part of the design considered here consists of about 5000 lines of VHDL code, resulting in 84000 gates (not counting on-board RAMs. Including RAMs there are 805 000 gates).

The basic function of the design is to transfer data packets as follows:

- read input from DSP lines into input buffer.
- transfer AAL2 packets to output queues and convert to ATM format
- read from output queues and forward to an ATM processing package.

5.1.2 Task. After completion of the design the designer left the company. Early tests indicated potential errors, which the new team could not easily locate. In addition, new requirements led to the decision for a re-design. In this situation a cooperation with the CVE team was started, with the goal,

- to locate and diagnose the errors in the design, if there are any,

- to provide a detailed documentation as the basis for the redesign,
- to specify the performance data in terms of burst data rates.

Our verification procedure can be characterized as a formally supported review, based on the gateprop tool.

Re-engineering method. gateprop was used for reverse engineering in the following steps:

1 Read and understand the design specification

2 Read (parts of) the VHDL code.

3 Write a property capturing one relevant function of the design.

4 Use gateprop to check whether the property holds.

5 If the property fails, analyse the counter-example: if it does not indicate a design bug, refine the property and go to 4.

6 If a design bug is found, report it and go to 3.

7 If the property holds, it documents a portion of the design. Repeat steps 3 to 6 until the design is fully covered by properties.

In the current project, the Debussy tool was used in addition to gateprop for analyzing the design. Debussy (by Novas) supports static interactive analysis of VHDL code such as tracing the drivers and loads of a signal, and visualizing the data flow structure. Whenever a counter-example shows that a certain signal behaves in some case different from the expectation, Debussy helps to quickly identify the particular situation when that occurs. Refining the property (step 5) means then that an assumption is added to exclude that situation.

5.1.3 Examples for a property. For illustration we choose (a slightly simplified version of) the property describing the control of the data transfer:

One transfer operation is started when

- the controlling finite state machine is in the start state, and
- there is enough space in the target queue.

To start the transfer, three control signals are set, and after 10 cycles the start state is reached again, so that the next target queue will be considered. This is formally expressed as follows:

```
theorem start_transfer;
for timepoints: t_s = t+2, t_end = t_s+11;
freeze:
  DEST_NO = act_op_queue @ t_s,
  SRC_NO  = bumx_trans_source_o @ t_s + 9;
assume:
  at t_s:   search_state = start;
  at t_s+8: space = '1';
  at t:     DEST_NO <= 8;
  at t:     SRC_NO <= 64;
  during [t, t_end]: noreset;
  <... others ...>
prove:
  during [t_s, t_s+9]: act_op_queue = DEST_NO;
  at t_s+ 9: bumx_trans_o = '1';
  at t_s+ 9: bumx_trans_source_o = SRC_NO;
  at t_s+ 9: bumx_trans_destination_o = DEST_NO;
  at t_s+10: search_state = start;
  at t_s+10: act_op_queue = next_q(DEST_NO);
end theorem;
```

5.1.4 Results. The complete circuit passed our VHDL front end in app. 10 minutes. The complete verification suite consists of 70 theorems, which are verified in app. 60 minutes.

Major results were:

- One deep design bug was revealed during the work. It occurs in the special case that due to a timeout, an incomplete packet is being read from the FIFO queue, and a new SDU packet is at the same time written to the FIFO. The bug occurs when the timing of these two events is within a window of two cycles no problem occurs if the writing starts before or after this window. Therefore it would be extremely difficult to reproduce, let alone discover this error by traditional simulation.

- The documentation specifies the detailed functionality and can be used both for performing the resign and for checking its correctness.

- The performance analysis showed that the current design can process at least 120 SDU packets in each 250 us cycle, which is almost twice the required rate.

5.2 Application Example: Complete Block-Level ASIC Verification

The ASIC considered in this section is developped by Siemens Mobile Communications. It serves as a 'number-cruncher' coprocessor in the uplink of a UMTS base station. An external client controls its operations, called tasks, via the control path of the ASIC, using parameter tables. The client is also responsible for starting and stopping the tasks. This is achieved by setting execution conditions, stored in execution tables, which are monitored by the control path [27].

The size of the coprocessor is more than 2 million logic gates, not counting on-chip RAM. Its clocking frequency is above 100MHz. The average data throughput lies in the range of several Gbit/s.

5.2.1 Verification Challenge and Approach. This project combines aspects of our "hard verification" scenario with the goal of complete early block-level verification.

The main challenge for the verification of the ASIC is its complicated task scheduling. It is dependent on many parameters. Most of the parameters are allowed to change every 10ms (the duration of one frame). This results in a case space of more than 10^{18} cases. Obviously, this case space could not be verified with classical approaches like simulation.

In addition to verifying the "difficult" blocks related to these scheduling function, gateprop was applied to almost every block of the ASIC. The exception was a few data-processing elements which seemed easy to validate by simulation.

The verification approach in our ASIC project combines block-level formal verification with system-level simulation. Working in a bottom-up way, each block was verified formally before the system level was even coded.

The verification team was headed by one experienced CVE verification engineer. All verification engineers had some background in formal methods and have been trained to use CVE. The major verification load was handled by this team, separating the concerns of designing and verifying. In addition, each designer underwent the same CVE training. Over the duration of the project the designers themselves increasingly started using the formal tool, e.g. adapting properties and running regressions.

In the verification process we can discern several phases. For each of them a close interaction of designer and verifier was practised:

Preparatory formalisation: using the informal specification, and interaction with the designer team, first properties were written even before block-level architecture is available. This required at least the entity description and saves some effort for the later phases - however in most cases the actual verification also required some knowledge of implementation details, such as the names of state variables.

Initial block verification: parallel to coding of a block properties were written, and as soon as the VHDL code could be successfully compiled, formal verification was started. Alternatively, the designer often ran a first simple simulation for a few standard cases before he handed the code over for verification. The latter option filters out some trivial bugs, but makes little difference from a global perspective. In this phase both trivial and complex bugs were found (and fixed), and as a result a formal block-level specification existed which covered the block's function completely, and truthfully with respect to the implementation.

Block-level regression: when a new HDL version was checked in the existing property suite was re-run, in some cases catching errors that were introduced by the change. This was often but not always possible without adapting the property suite.

Specification adaptation: as is common in a large innovative project as this, the system specification changed during the development duration. E.g., certain additional modes and flags were proposed by system engineering, and had to be incorporated into the already coded blocks. In this case properties had to be adapted in parallel with the code change, and re-checked. The complete regression suite was very helpful in these cases as it is by no means trivial to maintain the existing function while adding new ones.

Interface verification: Once two or more communicating blocks had been completely verified at the block level, the formal specification provided an excellent means to check their mutual interfaces. The ITL properties unequivocally describe the timing, handshake protocols and dependencies for each partner of an interface.

An example of such a situation is this: block A produces a result x, together with a "valid" flag x_valid. Unless this flag is set, the value is considered irrelevant. Now Block B, which consumes x, assumes that if x is not valid, it will not take the special value X'010', and acts on this assumption. None of the blocks is faulty by itself, but B's assumption is just not guaranteed by A.

Such dependencies were captured in ITL properties, and by carefully reviewing them, several bugs were discovered that resulted from two designers' deviating understanding of the informal specification.

5.2.2 Verifying the Control Path. Verifying control logic is an ideal application of property checking. Instead of generating sophisticated stimuli to check normal operation as well as the corner cases, properties cover all cases of an expected functionality at once. Moreover the specification of the control path is done using Finite State Machines. This directly allows to derive properties from the specification and check their validity in the RTL-description.

On the first synthesisable version of the code a large number of bugs were found. This is due to the fact that no simulation run preceded the formal verification, so a lot of simple bugs occurred. The advantage was that no effort was spent setting up a test bench at module level for the task controller.

The most important contribution of the formal verification was the detection of some difficult bugs. Often these were detected by the most general properties - such as mutual exclusion of events or a condition that always has to be met. While there was no error under most circumstances, the occurrence of several rare conditions at once caused a faulty behavior. Bugs of this kind would not have been caught by a simulation run.

5.2.3 Data Path Results. Our ASIC has many blocks with arithmetic functions. These contain more of the standard arithmetic such as complex addition and multiplication. Although arithmetic is typically problematic for formal verification tools, most blocks could be verified by CVE. In some cases we did encounter complexity problems, which, however, could always be alleviated by manual interaction, for example, by reducing the bit widths of the arithmetic operators. Typical design errors found in these blocks include wrong operand signs, wrong comparison operators ($<$ instead of $\leq$) and typical entity interface problems.

Note that property checking cannot completely replace simulation. For example, as stated before, designers usually maintain a test bench during the design process to be able to quickly simulate the basic functionality concurrently with coding and to immediately remove the most obvious design errors. Interestingly, after running these simulations, the designers very often felt sure that their designs were free of error, especially because their test benches operated the design at maximum load. A number of errors stayed undetected just because the designed system was simulated at "full-load" operating conditions, thereby simply disregarding certain slower input rates. These errors were, however quickly discovered by formal verification.

5.2.4 Overall Result. Block-level verification was completely covered by formal property checking. Comparing this to a traditional simulation-based approach, we have to consider several factors, as follows.

The total human effort for writing properties was about two person years. For a comparative discussion of such effort figures see the next section.

The computation time for a complete regression run has to be compared to the total simulator run time. On a basis of 40 blocks verified, we find that the sum of all verification run times is in the order of 50 CPU hours. However, fewer than 2 % of all theorems (10 out of a total of 660) account for more than 90% of this computation time. In other words 98% functional coverage is possible within only 5 hours. Given that simulator time is today one of the severely limiting factors in the ASIC quality assurance process, these figures show that formal property checking provides a very valuable progress here.

The quality achieved by block-level verification is probably the most important factor. It can be measured by the number of bugs which make it to the later stages of system simulation and emulation, or even to silicon. We have already reasonable evidence to claim that this number is cut down substantially. This claim is based on analyzing and classifying ca. 200 bugs discovered by property checking. This analysis showed that more than a third of these would have been missed by block-level as well as system-level simulation.

5.3 Productivity Statistics

It would be naive to believe that the benefits of formal verification come for free. From the experience of more than 20 ASIC projects we can quite well predict the effort required for completely verifying a given block: Per 1000 lines of VHDL code, the required effort ranges between 4 and 8 person days for verification, depending on such factors as

- the verification engineer's familiarity with the design,
- inherent complexity of the design,
- initial design quality, i.e. number of errors in the code.

When properties can be re-used because the design is very similar to one already formally verified, then this effort is naturally considerably reduced, typically to 1-3 days per 1000 LOC.

An other interesting parameter is the size of the property sets: compared to the HDL code size, they are regularly smaller by a factor of 1.5 to 5. This large diversity in compression factor is due to the wide variety of different scenarios - the factor is small when the design complexity comes from a complex specification such as complicated filter operations on the data path. It is high when there is a concise specification which abstracts from implementation details such as in a pipe-lined processor.

Note that the above figures apply for complete property suites, i.e. the properties form a complete case split of all situations faced by the design block. Thus, formal block verification is a well-controlled process with a clear termination criterion that produces blocks free from functional errors.

On the other hand, complete verification is not always the goal. Depending on the available resources, doing a "best effort", or just analyzing a few "critical" properties, or some aspect which escapes simulation, is also possible, and has often been done.

The above figure has to be compared to the total human effort for HDL coding on the one hand, and writing test benches on the other.

It is an often-quoted fact that total verification effort is typically 50-70% of the total project effort, while coding is in the 20-40% range. According to our experience, the effort for complete formal block verification is in the order of 50-80% of the coding effort.

Thus the total effort for block verification is reduced by 20-40% compared to a purely simulation-based approach. At the same time, a much higher block quality is obtained. Previously costly redesigns caused by errors in complex corner cases are now avoided. The savings in hardware and simulator licenses and the advantage of more reliable project schedules appear substantial but have not yet been quantified.

These estimates are based on adopting a two-heads policy, i.e. designer and verification engineer are distinct persons. Having the designer formally verify his block may save at least another 40% of the above formal block verification effort.

6. Summary

6.1 Achievements

Time to market and first-time-right silicon are the most important targets in today's ASIC development. We have demonstrated in many applications that property checking contributes significantly to meeting these targets.

Proof technology based on Boolean solvers has matured over many years to a high efficiency. We have given two examples of novel concepts for reasoning mechanisms exploiting higher-level structure, thus further extending the performance boundaries of formal methods.

6.2 Challenges and Perspectives

While formal property checking of design blocks is widely accepted as an efficient verification technology, several challenges remain. We briefly discuss the most important ones.

- Completeness proofs: While proving a single property provides the highest possible confidence in that property, the question "when have we written sufficiently many properties?" is of growing relevance. Based on the concept of case splitting, we do have criteria to check this completeness of a property set. However, these criteria are currently checked only manually and therefore, error-prone. Automating this check will be a significant step forward towards establishing formal proofs in a standard design flow. Compared to the traditional ways of measuring coverage, such a completeness proof is much more significant as it addresses functional rather than code coverage.
- Mixed-style verification: Simulation will remain an important tool in the overall verification flow. The division of work between formal and simulation-based techniques is currently decided ad hoc. A more systematic approach needs is called for, e.g. by stating properties and test-benches in a common style or language. The efforts to standardize a property specification language (PSL) may contribute to this unification.
- System on Chip: system level properties are out of the reach of today's technology - they require adequate abstraction techniques in order to avoid complexity explosion. Combining our efficient satisfiability-based techniques with abstract interpretation, and/or theorem provers looks like a promising approach.

Acknowledgments

The results reported here are to a large extent owed to the CVE team. Wolfram Büttner is acknowledged for his outstanding role in establishing this technology. Jörg Bormann, Dominik Stoffel, H.-Joachim Trylus and Görschwin Fey contributed important pieces of experience from the reported applications.

References

[1] Agrawal, M. and Thierauf, T. *The Boolean Isomorphism Problem.* IEEE Symposium on Foundations of Computer Science, pp. 422–430, 1996.

[2] The ATM Forum. *ATM Technology: The Foundation for Broadband Networks.* http://www.atmforum.com/

[3] Barrett, C. W. and Dill, D. L. and Levitt, J. R., *"Validity Checking for Combinations of Theories with Equality"*, Proc. FMCAD, pp. 187–201, 1996

[4] Barrett, C. W. and Dill, D. L. and Levitt, J. R., *"A Decision Procedure for Bitvector Arithmetic"*, Proc. DAC, pp. 522–527, 1998

[5] Biere, A. and Cimatti, A. and Clarke, E.M. and Zhu, Y. *Symbolic Model Checking Without BDDs* Proc. of Tools and Algorithms for the Analysis and Construction of Systems (TACAS'99), number 1579 in LNCS, pp 193–207, 1999.

[6] Bjørner, N. and Pichora, M. C., *"Deciding Fixed and Non-fixed Size Bit-vectors"*, Proc. TACAS, pp. 376–392, 1998

[7] Bormann, J., *"Productivity Figures for Complete Formal Block Verification"*, User Forum, DATE 2003

[8] Bormann, J., Spalinger, Ch., *"Formale Verifikation für Nicht-Formalisten"*, IT+TI 2/2001.

[9] Brinkmann, R. and Drechsler, R. *RTL-Datapath Verification Using Integer Linear Programming.* Proc. of ASP-DAC/VLSI Design 2002, January 07 - 11, 2002 Bangalore, India, pp. 741–746, 2002.

[10] Brinkmann, R. *Using Symmetry for Problem Reduction in Bounded-Model-Checking on the Register-Transfer-Level.* Proc. of SymCon'01 – Symmetry in Constraint Satisfaction Problems, CP'01 Post-Conference Workshop, 2001.

[11] Crawford, J. and Ginsberg, M.L. Eugene Luks, and Amitabha Roy. *"Symmetry-Breaking Predicates for Search Problems."* KR'96: Principles of Knowledge Representation and Reasoning, pp. 148–159. Morgan Kaufmann, 1996.

[12] Cyrluk, D. and Möller, M. O. and Rueß, H., *"An Efficient Decision Procedure for a Theory of Fixed-Sized Bitvectors with Composition and Extraction"*, Ulmer Informatik-Berichte, Fakultät für Informatik, Universität Ulm, 1996

[13] Cyrluk, D. and Möller, M. O. and Rueß, H., *"An Efficient Decision Procedure for the Theory of Fixed-Sized Bit-Vectors"*, Proc. CAV, pp. 60–71, 1997

[14] Markov, I.L. and Sakallah, K.A., and Aloul, F.A. and Ramani, A. *Solving Difficult SAT Instances in the Presence of Symmetry.* Proc. of ACM/IEEE Design Automation Conf., pp. 731–736, 2002.

[15] Ip, C.N. and Dill, D.L. *Better verification through symmetry.* Proc. of the 11th International Conference on Computer Hardware Description Languages and their Applications (CHDL'93), pp. 97–112, 1993.

[16] Johannsen, P. and Drechsler, R., *"Formal Verification on the RT-Level - Computing One-To-One Design Abstractions by Signal Width Reduction"*, Proc. VLSI, pp. 127–132, 2001

[17] Johannsen, P. and Drechsler, R., *"Speeding Up Verification of RTL Designs by Computing One-To-One Abstractions with Reduced Signal Widths"*, VLSI 2001 Post Conference Book, 2002, Kluwer Academic Publishers

[18] Johannsen, P., *"On Solving Systems of Bitvector Equations - An Efficient Decision Procedure for a Theory of Fixed-Size Bitvectors with Concatenation, Extraction and Negation"*, Siemens Corp., CT SE 4, 1999

[19] Johannsen, P., *"Reducing Bitvector Satisfiability Problems to Scale Down Design Sizes for RTL Property Checking"*, IEEE Proc. HLDVT, pp. 123–128, 2001

[20] Johannsen, P., *"Speeding Up Hardware Verification by Automated Data Path Scaling"*, PhD Thesis, Christian-Albrechts-University of Kiel, 2003

[21] Kravets, V.N. and Sakallah, K.A. *"Generalized Symmetries in Boolean Functions."* Proc. of ICCAD 2000, pp. 526–532, 2000.

[22] McKay, B.D. *Practical Graph Isomorphism.* Congressus Numerantium, volume 30, pp. 45–87, 1981.

[23] Moskewicz, M.W. and Madigan, C.F. and Zhao, Y. Zhang, L. and Malik, S. *Chaff: Engineering an Efficient SAT Solver.* Proc. of the 38th Design Automation Conference (DAC'01), pp. 530–535, 2001.

[24] Möller, M. O. and Rueß, H., *"Solving Bit-Vector Equations"*, Proc. FMCAD, pp. 36–48, 1998

[25] Paruthi, V. and Kuehlmann, A. *"Equivalence Checking Combining a Structural SAT-Solver, BDDs, and Simulation."* Proc. of the IEEE International Conference On Computer Design: VLSI In Computers and Processors (ICCD '00), pp. 459–464, 2000.

[26] Plump, D. *"Handbook of Graph Grammars and Computing by Graph Transformation"*, volume 2, chapter Term graph rewriting. World Scientific, 1998. Ehrig, H. and Engels, G. and Kreowski, H.-J.and Rozenberg, G.,editors.

[27] Winkelmann, K., Trylus, J., Stoffel, D., Fey, G., *"Cost-Efficient Formal Block Verification for ASIC Design"*, in: Drechsler, R., Metho-

den und Beschreibungssprachen zur Modellierung und Verifikation von Schaltungen und Systemen, 6. GI/ITG/GMM-Workshop Modellierung und Verifikation, Shaker-Verlag 2003.

Chapter 5

ASSERTION-BASED VERIFICATION

Property Specification

Claudionor Nunes Coelho Jr.
Computer Science Department, Universidade Federal de Minas Gerais, Belo Horizonte, MG - Brazil
coelho@dcc.ufmg.br

Harry D. Foster
Jasper Design Automation, Inc, Mountain View, CA - USA
harry@jasper-da.com

Abstract Assertion-based verification – that is, user-specified properties combined with simulation, formal techniques, and even synthesis – is likely to be the next revolution in hardware design and verification. This chapter explores a verification break-through prompted by multi-level specification and assertion verification techniques. The emerging Accellera PSL formal property language, as well as the Accellera Open Verification Library standards and the important roles they will play in future assertion-based verification flows are discussed.

Keywords: assertion, assumption, constraint, dynamic verification, formal verification, restriction, static verification, property, specification, synthesis

1. Introduction

As formal research matures and approaches a level of sophistication required by industry (beyond the bounds of research and early adopters), we must take steps to ensure a successful transfer (scaling) to this more demanding level. One step is to fundamentally change design methodologies such that we move from ambiguous natural language forms of specification to forms that are mathematically precise and verifiable. Furthermore, these languages must lend themselves to automation. For-

R. Drechsler (ed.), Advanced Formal Verification, 167-204.

mal *property specification* is the key ingredient in this methodological change. The end result is higher design quality through:

- *improved understanding of the design space* – resulting from the engineer's intimate analysis of the requirements, which often uncovers design deficiencies prior to RTL implementation
- *improved communication of design intent* among multiple stakeholders in the design process
- *improved verification quality* through the adoption of assertion-based verification techniques

Although the need for methodological change is clear, transitioning formal verification technology into an industry design environment has been limited by a lack of methodology guidelines for effective use.

Property specification (that is, assertions, constraints, and functional coverage) is fundamental to an assertion-based verification platform. Once specified, properties enable the following components, which may be included in an assertion-based verification platform:

- *verifiable testplans* through property specification (for example, executable *functional coverage models*, which help answer the question *"what functionality has not been exercised?"*)
- *exhaustive* and *semi-exhaustive* static formal property checking technology (for example, model checking and bounded-model checking)
- *dynamic property checking* technology (for example, monitoring assertions in simulation) for improved observability that reduces the time involved in debug
- *hardware verification languages* (HVLs) for testbench generation that leverage property specification to define expected input (constraints) and output (assertions) behavior
- *constraint-driven stimulus generation* based on interface properties that target block-level designs
- *assertion property synthesis* to address silicon observability challenges during chip bring-up in the lab, as well as operational error detection required for high availability (HA) class systems

Dynamic verification, for the foreseeable future, will remain a critical component of an assertion-based methodology for two reasons: (1) the inherent capacity limitations of today.s formal technology, and (2) the unanswered question – *"Is the formal specification complete?"* - In other words, formal verification can help us answer questions about the design, provided we ask the questions in terms of a property. However, if we neglect to ask a question, then we might have a false sense of security that the design is correct. Similarly, dynamic verification is inherently incomplete. However, during random stimulus generation, it can occasionally be useful for identifying design errors that address questions we would have never thought to ask.due to an incomplete specification.

In this chapter, we discuss the important role that property specification plays in an assertion-based verification flow. We begin with a discussion of a property specification framework, we compare and contrast basic linear and branching time temporal logic, and we introduce the Accellera PSL property specification language [2]. Next, we introduce the idea of creating a library of assertion monitors that can be used in various forms of verification. We then focus our discussion on assertions in simulation, followed by a discussion of assertions in formal verification. Next, we explore how assertions can actually be synthesized to address silicon observability challenges during lab bring-up, as well as operational error detection required for high availability (HA) class systems. Finally, we demonstrate examples of property and assertion specification and close with a discussion of future directions with property specification and assertion-based verification.

1.1 Specifying properties

Informally, a *property* is a description of design intent. When studying properties, it is generally easier to view their composition as three distinct layers:

- the *Boolean layer*, which is comprised of Boolean expressions (for example, Verilog or VHDL expressions)
- the *temporal layer*, which describes the relationship of Boolean expressions over time
- the *verification layer, which describes how to use a property during verification*

Defining (or partitioning) a property in terms of the abstract layer view enables us to dissect and discuss various aspects of properties. However, it is quite simple to express design properties, and the three-layer view is merely a way to explain concepts. We do not intend for our use of this vehicle to convey a sense that the actual language syntax is complex.

To aid in studying property concepts, all examples in the following sections are presented using the Accellera PSL property specification language, unless otherwise noted.

Boolean layer: A property's Boolean layer is comprised of Boolean expressions composed of variables within the design model. For example, if we state that *"signal* `en1` *and signal* `en2` *are mutually exclusive,"* then the Boolean layer description representing this property could be expressed in Verilog as shown in Example 5.1.

Example 5.1 Boolean layer expressed in Verilog

```
!(en1 & en2) // enables are mutually exclusive
```

Temporal layer: A property's temporal layer permits to describe the Boolean expressions. relationships to each other over time. Thus, all time ambiguities associated with a property are removed. For example, if signal en1 and signal en2 are always mutually exclusive (that is, for all time), then a temporal operator could be added to the Boolean expression to state precisely this. Temporal operators allow us to specify precisely when the Boolean expression must hold. Example 5.2 demonstrates this point using the PSL temporal operator *always* combined with a Verilog Boolean expression.

Example 5.2 A property's temporal layer expressed in PSL

```
always !(en1 & en2) // enables are mutually exclusive
```

Together, the Boolean and temporal layers form the foundation of a property.

Verification layer: While a property.s Boolean and temporal layers describe general behavior, they do not state how the property should be used during verification. In other words, should the property be asserted, and thus checked? Or should the property be assumed as a constraint? Or should the property be used to specify an event used to gather functional coverage information during simulation? Thus, it is the third layer of a property, which is the *verification layer*, that states how the property is to be used.

Look again at the property *"signal* `en1` *and signal* `en2` *are mutually exclusive."* Example 5.3 shows this property with the PSL assert

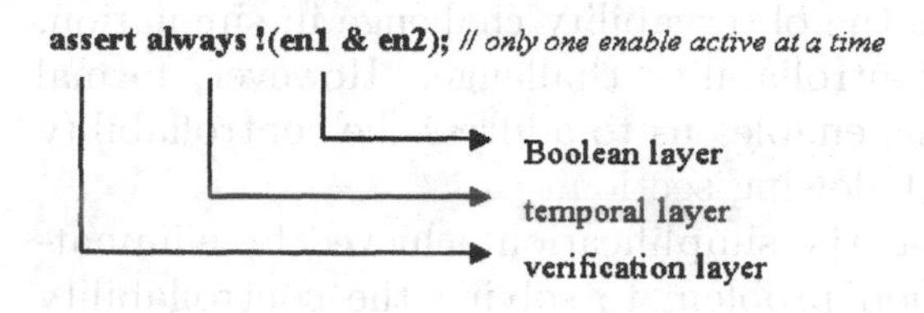

directive. This states that the property is to be treated as an assertion during verification.

Example 5.3 A property's verification layer expressed in PSL.

1.2 Observability and controllability

Fundamental to the discussion of assertion-based verification is understanding the concepts of controllability and observability. *Controllability* refers to the ability to stimulate a specific line of code or structure within the design. Note that, while in theory a simulation testbench has high controllability of the input bus of its device under verification, it can have low controllability of an internal point. *Observability*, in contrast, refers to the ability to observe the effects of a specific internal, stimulated line of code or structure. Thus, a testbench generally offers limited observability, if it only observes what is on the external ports of the device or model, because all the internal signals and structures are often hidden from the testbench. To identify a design error using the testbench approach, the following conditions must hold:

- proper input stimulus must be generated to activate (that is, sensitize) a bug
- proper input stimulus must be generated to propagate all effects resulting from the bug to an output port

It is possible, however, to set up a condition where the input stimulus activates a design error that does not propagate to an observable output port. In these cases, the first condition cited above applies; however, the second condition is absent.

A benefit of assertions embedded in the design model is that they increase *observability*. In this way, the verification environment no longer depends on the second condition listed above to identify bugs. Thus, any improper or unexpected behavior can be caught closer to the source of the bug, in terms of both time and location in the design intent.

While assertions help solve the observability challenge in simulation, they do not help with the controllability challenge. However, formal property checking of assertions enables us to address the controllability challenge as discussed in the following section.

The reader should note that the simplification achieved by eliminating or reducing the verification problem to solving the controllability problem in a design does not make the verification problem any easier, as selecting the right pattern to sensitize an error condition can be as hard as sensitizing the design to propagate the erroneous condition to the output pins of the design.

1.3 Formal property checking framework

In this section, our goal is to introduce the basic elements of formal property checking and in so doing, convey a sense of both its inherent power and limitations. Steps required to perform formal property checking (for example, *model checking*) include:

- compile a formal model of the design
- create a precise and unambiguous specification
- apply an automated and efficient proof algorithm

Each of these steps is briefly discussed below.

Compile a formal model: In the first step of the formal property checking process, we create a formal model of the design by compiling a non-ambiguous description of the model (usually a synthesizable subset of a hardware description language, such as a Verilog RTL model) into a form accepted by the property checker. For the purpose of our discussion, we consider hardware designs as a set of finite state concurrent systems. For example, the value of the *current state* of the system can be determined at a particular point in time by examining all state-elements of the system. The *next state* of the system can be computed as a function of the system's current state value and design input values. This function is called a transition function. In formal verification, we can conveniently represent a current state - next state pair as a *transition relation* of the system. For example, (s_i,s_{i+1}) is a transition relation ,where s_i represents a current state of the system, and s_{i+1} represents one next state possibility directly reachable from s_i.

In a formal representation of a design, we use *state* to indicate any variable retaining its value over time. In that broader sense, a state can be considered to be the inputs of the design, the variables from the control path representing the finite-state machines of the design and the variables from the datapath representing stored results from operations, such as the results from ALU operations. In Example 5.4, we demonstrate a state representation for a simple design containing two inputs and three internal variables.

Example 5.4 State representation in a design: in an RTL design with inputs *Reset* and *Clock*, and variables *State, AluOut,* and *Flags*; a state for this system is represented by all possible combinations of the tuple $s = <Reset, Clock, State, AluOut, Flags>$. A transition from state s_i to state s_{i+1} is represented by all possible sequences of tuples ¡*Reset, Clock, State, AluOut, Flags*¿.

A *path* at state is an infinite sequence of states $\pi = s_0 s_1 s_2 ...$, which represents a forward progression of time and a succession of states. Note that a simulation trace is one example of a path. A set of paths represents the *behavior* of the system. Hence, a formal model can be created by compiling a synthesizable model of the design into a state transition graph structure, referred to as a *Kripke structure* [10].

A Kripke structure M is a four tuple $M = (S, S_0, R, L)$, which consists of:

- S a finite set of *states*
- S_0 a set of *initial states*, where $S_0 \subseteq S$
- $R \subseteq S \times S$ a transition relation, where for every state $s \in S$. There is a state $s' \in S$ such that $(s, s') \in R$
- $L : s \rightarrow 2^{AP}$, where L is a function that labels each state with a set of atomic propositions (AP) that are true at that particular state

A Kripke structure models the design using a graph, where a node represents a state, and an edge represent transition between states. Atomic propositions map Boolean variables (and their negation) into the formal model of a design, represented by S, S_0, and R. For any atomic proposition p, if $p \in L(s)$, we say p is true (holds) in s. Similarly, if $p \notin L(s)$, we say p is false (does not hold) in s. By analyzing when an atomic proposition holds in a state, we can verify if a property is true or false.

Example 5.5 Consider a sequential 2-bit counter with counting variable *Count[1:0]* and an input pin *Clock*. We can build a Kripke structure based on the states formed by the tuple ¡*Clock, Count[1], Count[0]*¿. The Kripke structure is presented in Figure 5.1. For this Kripke structure, *Count[1]* holds for the states s_3, s_4, s_5, and s_6.

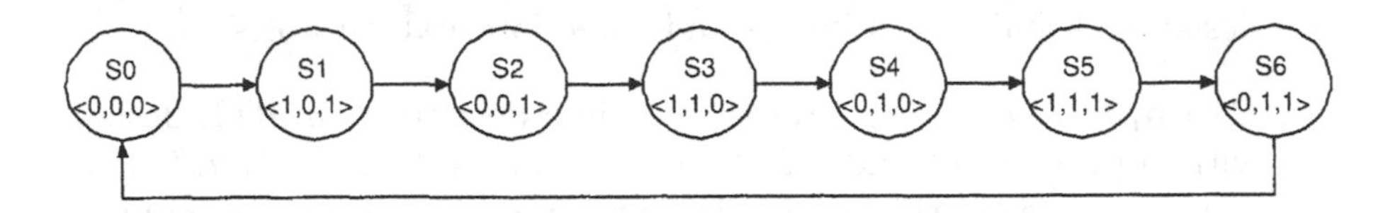

Figure 5.1. Kripke structure for 2-bit counter.

Create a formal specification: In the next step of formal property checking, properties are specified as assertions of the design that we wish to verify. Informally, a property describes *design intent.* More formally, a property is defined as follows:

Definition 1: property – a collection of logical and temporal relationships between and among subordinate Boolean expressions, sequential expressions, and other properties that in aggregate represent a set of behavior (that is, a path).

A safety property is defined as follows:

Definition 2: safety property – a property that specifies an invariant over the states in a design. The invariant is not necessarily limited to a single cycle, but it is bounded in time. Loosely speaking, a safety property claims that something bad does not happen. More formally, a safety property is a property for which any path violating the property has a finite prefix such that every extension of the prefix violates the property.

For example, the property, *the signals* `wr_en` *and* `rd_en` *are mutually exclusive* and *whenever signal* `req` *is asserted, signal* `ack` *is asserted within 3 cycles* are safety properties.

A liveness property is defined as follows:

Definition 3: liveness property – a property that specifies an eventuality that is unbounded in time. Loosely speaking, a liveness property claims that *something good* eventually happens. More formally, a liveness property is a property for which any finite path can be extended to a path satisfying the property.

For example, the property *whenever signal* `req` *is asserted, signal* `ack` *is asserted some time in the future* is a liveness property.

Finally, a fairness property is defined as follows:

Definition 4: fairness property - property that specifies that some condition will occur infinitely often. More formally, a fairness property is a property for which any infinite path will contain infinitely many fair states, or states that satisfy the fairness condition of the property.

Underlying many property languages is a formalism known as *propositional temporal* logics, which allows us to reason about sequences of transitions between states. Two formalisms for describing sequence propositions are branching-time temporal logic [4] and linear-time temporal logic [13]. CTL is an example of branching-time logic. The temporal operators of this formalism allow us to reason about all paths originating from a given state. Whereas in the case of LTL (a linear-time temporal logic), the temporal operators allow us to reason about events along a single computation path.

Applying a proof algorithm Once we have created a formal model representing the design and a formal specification precisely describing a property that we wish to verify, our next step is to apply an automated proof algorithm. For example, given a formal model of a design described as a Kripke structure $M = (S, S_0, R. L)$, and a temporal logic formula f expressing some desired property of the design, the problem of proving the correctness of f involves finding the set of all states in S that satisfy f.

Note that the formal model satisfies the specification *if and only if* all initial states (that is, $\forall s_i \in S_n$) are in the set of the states that satisfies f. A procedure for determining a set of states satisfying f is informally shown in Figure 5.2.

The illustrated proof algorithm we use is known as reachability analysis using *image computation*. This algorithm is the basic algorithm for proving that a temporal property f is valid.

The algorithm begins with a set of initial states S_0, as shown in Figure 5.2. Using the transition relation R, as previously discussed, we calculate within one step (that is, a tick of the clock) all reachable states from S_0. This calculation process is referred to as *image computation.* The new set of reachable states is S_1 in our example. We iterate on this process, generating a new set of reachable states at each step that grows monotonically, until no new reachable states can be added to the new set (that is, a *fixed-point* occurs when $S_k == S_{k+1}$).

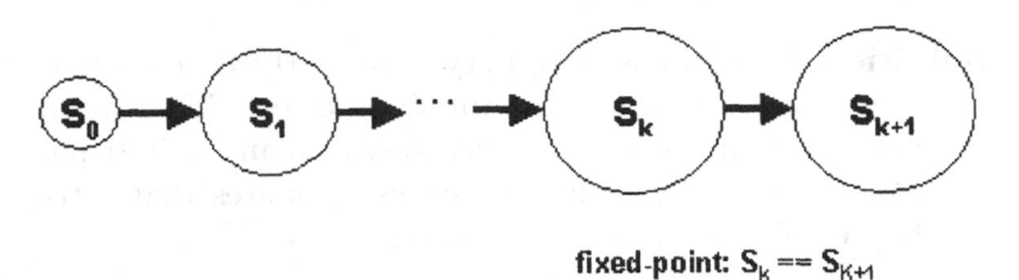

Figure 5.2. Calculating reachable states

For example, if the temporal formula f describes a safety property, we can validate that f holds on each new state calculated during the image computation step.

Proof results For this fixed-point proof algorithm, one of three possible results occurs:

- **Pass.** The process reaches a *fixed-point*, and the formula f holds on all reachable states. Hence, the verification is done (that is, the design is valid for this property).
- **Fail.** The process has yet to reach a fixed-point, and the formula f was determined not to hold on a particular state s_i, which was calculated during the search. Hence, a *counter-example* (that is, a path $\pi = s_0 s_1 s_2 ..., s_j$) can be calculated back from the bad state s_j to an initial state. This counter-example is then used to debug the problem.
- **Undecided.** The process aborts prior to reaching a fixed-point due to a condition known as *state-explosion* (that is, there are too many states for the proof engine to represent in memory). In the following section, we discuss a few techniques that address the state-explosion problem.

Formal property checking tools use a number of different proof algorithms. A detailed discussion of these specific *proof algorithms, creating formal models,* and *temporal logics* is beyond the scope of this chapter. For in-depth discussions on these subjects, we suggest [5] and [11].

2. Assertion Specification

In this section, we discuss branching-time and linear-time temporal logic as a foundation for our introduction to the Accellera Property Specification Language (PSL).

2.1 Temporal logic

Temporal logic enables us to reason about systems in a very simple way by hiding time relationships between Boolean formulas. For example, instead of writing $\forall t.!(en1(t)\&en2(t))$, in which time t is explicit in the temporal formula, we may simply write `always !(en1 & en2)`, which states that `en1` and `en2` should not hold at the same time.

In this chapter, we assume a temporal logic in which existential and universal quantifiers can be applied only to time, which is called *propositional temporal logic* in the literature. Even though quantification can be applied only to the time variable, when we prove temporal formulas we will be proving formulas over paths of execution ($\pi = s_0 s_1 s_2 ...$). In the formula $\forall t.!(en1(t)\&en2(t))$, for example, proving this property for path π implies that we will be proving the conditions $!(en1(s_0)\&en2(s_0))$, $!(en1(s_1)\&en2(s_1))$, ...

When we prove a temporal logic property, we may assume that a state has only one successor, in which case the property is proven on a given path or trace of execution. This logic is called linear time temporal logic. We may also assume that each state may have several successors, in which case we have to prove the property on a set of paths. The latter case is usually represented as an infinite tree and it is suitable for representing computations. This logic is called branching time temporal logic.

To show how these two types of logic differ, we first introduce CTL*, which contains both CTL and LTL, and plays essential roles in formal hardware verification for branching time temporal logic and linear time temporal logic, respectively.

CTL* contains operators for reasoning about paths of computation, such as the operators **G** (always), **F** (eventually), **U** (until) and **X** (next) and operators for reasoning on branching paths of execution, that is, the operators **A** (for all paths of execution) and **E** (for any path of execution). In addition to these quantifiers, any Boolean composition of CTL* formulas are CTL* formulas as well.

For any temporal formulas p and q, the temporal formula **G** p means that p always holds, or the temporal formula p holds for all states of a path π. The temporal formula **F** p means that p holds for some future state of a path π. The temporal formula p **U** q means that the temporal formula p will be valid in all states of a path π until q is true in some future state of π.

Path quantifiers behave as described previously. The temporal formula **A** p means that for all paths π starting with the current state, p will be valid. The temporal formula **E** p means that there is a path π starting with the current state for which p is valid.

As seen in the previous paragraphs, CTL* can be separated into *state formulas* and *path formulas.* Any atomic proposition or Boolean formula over *state formulas* is a *state formula.* In addition, existential quantification over path formulas (**E** *path*, where **E** is the existential quantifier and *path* is a path formula) is also a *state formula.*

Any *state formula* is a *path formula,* as are *path formulas* Boolean composition of *path formulas.* In addition, *path formulas* can be composed using the temporal operators **X** *path* and $path_1$ **U** $path_2$.

The reader should note that the **F** operator can be thought of as an alias for the unary form of the *until* operator (**F** p = **true** **U** p), and that **G** and **F** are dual, that is, **G** p is equivalent to !**F** !p. The rationale behind the first alias is that eventually p is equivalent to waiting vacuously until p occurs; and the rationale behind the second formula is that stating that p is always true is equivalent to saying that it is not the case that !p will be true in the future. Similarly, it is not difficult to show that **E** and **A** are dual.

Note that in CTL* we do not make any restriction on the order in which temporal and branching operators appear in a valid formula. As a result, **FG** p and **AG** p are valid CTL* formulas. The first formula states that eventually p will be true forever. The second formula states that for all paths starting from the current state, p always will be true.

Valid CTL* formulas can be proven valid (or not) in formal models of a design. Let f be a CTL* temporal formula expressing some desired behavior, and let M be a formal model of a design described as a Kripke structure (S, S_0, R, L). The problem of proving the correctness of f for the model M involves finding the set of all states in S that satisfy f.

$$\{\forall s \in S \text{ s.t. } M, s \models f\}$$

where $M, s \models f$ means the property represented by the temporal formula f holds at state s of model M. Note that the formal model satisfies the specification *if and only if* $\{\forall s \in S_0\} \subseteq \{\forall s \in S \text{ s.t. } M, s \models f\}$.

Now that we have presented CTL* to the reader, we can restrict this logic to CTL and LTL.

- **CTL:** a CTL formula is a CTL* formula beginning by a branch quantifier (**A** and **E**), restricting that temporal operators (**F**, **G**, **U**, and **X**) be preceded by branch quantifiers For example, the formula **AG** p is a valid CTL formula, but **FG** p is not.

- **LTL:** an LTL formula is the subset of CTL* formulas obtained by simply restricting the valid formulas to path formulas. For example, the formula **FG** p is a valid LTL formula, but **AG** p is not.

If we consider an implicit universal quantifier for all paths in front of an LTL formula, we can see that certain behaviors, such as **A**(**GF** f) cannot be represented by CTL, although it is a valid LTL formula. Similarly **AG**(**EF** f) is a valid CTL formula, but not a valid LTL formula. While the first formula states that for all states of all paths, eventually f will be valid (a fairness constraint), the second formula states that for all branches of all states, at least in one of the paths eventually will have a valid f.

It is beyond the scope of this chapter to discuss thoroughly the semantics of CTL*, CTL, and LTL. Our goal is to give a short introduction on this subject to support the remainder of this chapter. We refer the reader to [11] to a more complete definition of these logics' semantics and complexities.

Although writing temporal formulas is much easier than writing their formulas with explicit time quantifiers, these temporal formulas can still cause major problems during the verification procedure, leading to false positive results due to incorrect formula specification.

2.2 Property Specification Language (PSL)

The Accellera Property Specification Language (PSL) is an ideal language for specifying design intent in either linear-time temporal logic or an optional branching-time temporal logic. In Section 1.1, we presented a segmented layering concept as a matter of convenience to describe a property language. Similarly, the PSL language definition is segmented into layers: *Boolean, temporal, modeling,* and *verification.* The temporal layer supports either the LTL linear-time temporal logic or CTL branching-time temporal logic operators. In this section, we consider only the linear-time temporal logic component known as the PSL *foundation language* (FL). For a more complete definition, see [2].

2.2.1 Boolean layer. At the Boolean layer, a PSL specification references signals and variables within an HDL description (for example, Verilog or VHDL). Hence, the underlying HDL syntax and semantics for Boolean expressions ensure semantic consistency between the property specification and the HDL model.

2.2.2 Temporal layer.

Sequences: Sequences of Boolean conditions that occur at successive clock cycles can be described succinctly using *Sequential Extended Regular Expressions* (SEREs). Sequences and SEREs can be constructed as follows (where b is a Boolean expression):

- **b** – a Boolean expression is a SERE in its simplest form
- **{SERE}** – a sequence constructed by a SERE
- **SERE ; SERE** – a SERE constructed by concatenating two SEREs
- **{sequence — sequence}** – a sequence describing alternatives
- **{sequence & sequence}** – a sequence describing parallel non-length matching sequences (that is, two sequences, both hold at the current cycle, regardless of whether they complete in the same cycle or in different cycles)
- **{sequence & sequence}** – a sequence describing parallel length matching sequences (that is, two sequences, both hold at the current cycle, and both complete in the same cycle)

PSL provides various repetition operators ([ ]) that concisely describe repeated concatenation of the same SERE. For example, given the SERE r and a Boolean b:

- `r[*m:n]` – a sequence of n contiguous occurrences of `r`
- `b[=m:n]` – any sequence containing n occurrences of `b`
- `b[->m:n]` – any sequence ending in the nth occurrence of `b`
- `r[*]` – zero or more occurrences: `r[*0:inf]`
- `r[+]` – at least a single occurrence: `r ; r[*]`

The repeat range $m : n$ can be replaced by a single constant n (for example, `[*2]`). In addition, an unbounded range could be expressed as `[*0:inf]`, where the keyword `inf` represents infinity. Note that the + and * qualifiers may stand alone, without a preceding SERE, in which case they represent the advancement in time within a range of cycles (for example, {`a;[*2];b`} = {`a;'TRUE[*2];b`}).

Properties: PSL supports all the standard LTL operations. In addition, more readable operators are defined in terms of the base operators. For example, given the PSL temporal formulas `f`,`f1`, `f2`, a few of the common PSL FL operators include:

- `!f` – `f` does not hold
- `f1 & f2` – `f1` and `f2` both hold
- `f1 | f2` – `f1` or `f2` or both hold
- `f1 -> f2` – `f1` implies `f2`
- `f1 <-> f2` – `f1 -> f2` *and* `f2 -> f1`
- `always` *f* – f holds in every cycle: **G** f
- `never` *f* – f does not hold in any cycle: **G** !f
- `next` *f* – f holds in the next cycle, if any: **X** f
- `next!` *f* – f holds in the next cycle: **X**!f
- *f1* `until` *f2* – f1 holds until f2 holds, if ever: f1 **W** f2
- *f1* `until!` *f2* – f1 holds until f2 eventually holds: [f1 **U** f2]
- *f1* `before` *f2* – f1 holds before f2 holds
- `within`(*r1*, *b*)*r2* – r2 occurs after r1 and before b
- `eventually!` *f* – an f holds in some future cycle: **F** f

Notice the `eventually!` operator, which is referred to as a strong operator due to the exclamation mark (!). PSL supports both strong and weak operator forms. A *strong form* requires the terminating condition to eventually occur, while the *weak form* makes no requirement about the terminating condition. For example, the strong and weak forms of *"busy shall be asserted until done is asserted"* are (`busy until! done`) and (busy until done), respectively. The strong form (that is, tt until!) states that `busy` shall be asserted until `done` is asserted, and that done shall eventually be asserted. The weak form (that is, `until`) states that `busy` shall be asserted until `done` is asserted, and that if done is never asserted, then `busy` shall stay asserted forever. Note that a property that uses a non-negated strong operator is a liveness property, while one that uses only non-negated weak operators is a safety property.

PSL also supports operators that build complex properties out of SEREs:

- `{r1} |-> {r2}` - r2 starts in the last cycle of r1 (overlap)
- `{r1} |=> {r2}` - r2 starts in the first cycle after r1
- `{r} (f)` - f holds in the last cycle of r

Declarations: PSL allows us to define named property and sequence declarations with optional arguments, which facilitates reuse. These parameterized declarations can be referenced by name and instantiated in multiple places in designs with unique argument values.

For example, we could specify the property that $en1$ and $en2$ are mutually exclusive as follows:

Example 5.6 PSL property declaration

```
property mutex (boolean clk, a, b) =
  always !(a & b )@(posedge clk);
```

2.2.3 Verification layer.

Directives: The verification layer provides *directives* that tell the verification tools what to do with the specified properties. Directives specify whether a given property is expected to hold (that is, an assertion) or assumed to hold (that is, an assumption) as a constraint. Similarly, other directives specify whether verification should exclude situations in which a given sequence occurs (that is, a restriction) or ensure that other sequences are encountered during verification (that is, a functional coverage specification). The PSL verification directives are:

- **assert**
- **assume**
- **assume_guarantee**
- **restrict**
- **restrict_guarantee**
- **cover**
- **fairness** and **strong fairness**

Clock declaration: PSL provides a means for specifying a *default clock* expression, which enables us to define multiple properties or sequences without explicitly specifying a clock. For example:

Example 5.7 PSL default clock declaration

```
default clock = (posedge clk);
assert always !(en1 & en2);
```

Alternatively, we can explicitly associate a clock with a property or sequence as follows:

Example 5.8 PSL explicit clock declaration

```
assert always !(en1 & en2) @(posedge clk);
```

3. Assertion libraries

There is a large class of properties that can be reused, which ranges from a higher-level common interface protocol down to lower-level RTL implementation properties. For example, using PSL, we could express a property that signals *en1* and *en2* are always mutually exclusive as follows:

Example 5.9 PSL property for mutually exclusive signals

```
property mutex =
    always (reset_n != 1'b0 || !(en1 & en2))@(posedge clk);

assert mutex;
```

However, we could encapsulate the properties into a parameterized template or module, which can be pre-verified and then reused. The Accellera Open Verification Library (OVL) [7] is an example of an assertion library ideal for use in an industry setting. For example, we could express the assertion that a and b are always mutually exclusive using an OVL `assert_always` Verilog module as follows:

Example 5.10 OVL assertion monitor for mutually exclusive signals

```
assert_always mutex (clk, reset_n, !(en1 & en2));
```

The OVL assertion monitors provide many systematic elements for an effective assertion-based verification methodology that are typically not addressed by general property languages. For instance, the OVL incorporates a consistent and systematic means of specifying RT-level implementation assertions structurally through a set of concurrent assertion monitors. These monitors provide designers with a module, which guides them to express a broad class of assertions. In addition, these monitors address methodology considerations by providing uniformity and predictability within an assertion-based verification flow and encapsulating the following features:

- unified and systematic method of reporting that can be customized per project
- common mechanism for enabling and disabling assertions during the verification process
- systematic method of filtering the reporting of a specific assertion violation by limiting the firing report to a configured amount

One particularly compelling aspect of the OVL is that it does not require a compilation step to take advantage of assertion specification in the RTL source. Furthermore, the assertion library has proven its value in an industrial setting since the library is written in standard IEEE-1364 Verilog and IEEE-1076 VHDL, and it works with any commercial simulator. This means that IP containing assertions can be delivered to customers without delivering any additional tools for preprocessing the assertion into simulation monitors.

4. Assertion simulation

An assertion-based methodology leverages the assertion specification between multiple verification techniques, for example static formal property checking and dynamic simulation-based verification. Although formal verification has proven its worth in an industry setting, for the foreseeable future simulation will play a critical role in the design verification flow. In this section, we discuss the role of simulation in an assertion-based methodology.

Currently, various techniques support assertions within a simulation environment; for example, adopting a library of assertion monitors (as previously discussed), including native simulator support for assertion languages (like PSL), or translating an assertion specification into a checker or monitor that is integrated into the simulation environment [3]. Figure 5.3 shows an environment for translating a formal property language (such as PSL) into a simulation monitor. The user provides a design to be verified, as well as formal specifications and a set of test programs generated either manually or automatically. During simulation, the translated checker reports any property violations.

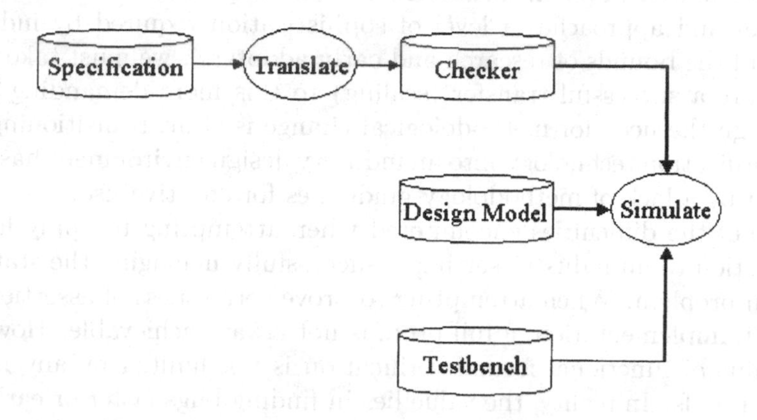

Figure 5.3. Assertion checker generation from specification

Assertions in simulation have been used by many prominent companies, including: Cisco Systems, Inc., Digital Equipment Corporation, Hewlett-Packard Company, IBM Corporation, Intel Corporation, LSI Logic Corporation, Motorola, Inc., and Silicon Graphics, Inc.

Designers from these companies describe their success with methodologies that incorporate assertions as follows:

- **34%** of all bugs uncovered with simulation were found by assertions on DEC Alpha 21164 project [9]
- **25%** of all bugs uncovered with simulation were found by assertions on DEC Alpha 21264 project – The DEC 21264 Microprocessor [14]
- **25%** of all bugs uncovered with simulation were found by assertions on Cyrix M3 (p2) project [6]

From these published results, a common theme emerges: when designers use assertions as a part of the verification methodology, they are able to detect a significant percentage of design failures. Thus, assertions not only enhance a verification methodology; they are an integral component. Finally, one significant aspect of specifying assertions is, as previously stated, that assertions improve observability. This results in a significant reduction in simulation debug time – a savings of up to 50% as reported by [3].

5. Assertions and formal verification

As we stated in our introduction to this chapter, as formal research matures and approaches a level of sophistication required by industry (beyond the bounds of research and early adopters), we must take steps to ensure a successful transfer (scaling) to this more demanding level. Although the need for methodological change is clear, transitioning formal verification technology into an industry design environment has been limited by a lack of methodology guidelines for effective use.

One of the difficulties encountered when attempting to apply formal verification to an industry setting is successfully managing the state explosion problem. When attempting to prove correctness of assertions on an RTL implementation, a full proof is not always achievable. However, the value of functional formal verification is not limited by any means to full proofs. In reality, the value lies in finding bugs faster or earlier in the design cycle and finding difficult bugs missed by traditional simulation approaches, which in turn increases confidence in the correctness of the design while decreasing time to market. If our goal is an exhaustive proof, verification for assurance versus verification for bug hunting, then typically some form of abstraction is typically required on portions of the design (for example, counters and memories).

Another difficulty encountered when attempting to apply formal verification to an industry setting is the methodological requirement for accurately specifying environment constraints. The formal verification engine uses the constraints to limit the exhaustive search to a valid set of legal behaviors. Note that the work used to create block-level environmental constraints for a formal verification engine can often be re-used as block-level interface assertions during full-chip and system simulation. Hence, there is a return on investment for formally specifying block-level interface properties that include, as previously stated, improved understanding of the design space, improved communication of design intent, and improved verification quality.

5.1 Handling complexity

In this section, we discuss techniques typically used to handle the state explosion problem when proving properties on industrial RTL models.

Choose appropriate RTL. The first step in handling complexity is to initially choose the right level of RTL on which to apply formal. For example, RTL contained in control-intensive logic is better suited for formal property checking than RTL that models data path logic. Size of the RTL component (in terms of state directly related to a property) must be considered. Other factors that influence the RTL selection are design-related. For example, not every RTL component (that is, module, block, or unit) is a good candidate for stand-alone verification. Interesting properties may require more logic to be included beyond our selected RTL component. This can be problematic since many internal interfaces are rarely documented. Furthermore, the additional logic not included with our RTL component that we wish to verify may be too complex to model as environment constraints. Nonetheless, if we choose the appropriate RTL wisely, we can have a high degree of success at formally verifying properties on RTL components.

Property decomposition. We recommend that complex sequential assertions be split into simpler assertions. For example, break a req-ack handshake down into its component elements (arcs on a timing diagram). This think static rather than dynamic approach works well for formal proofs.

Compositional reasoning. One technique to cope with the state explosion problem is to partition a large unverifiable component into smaller, independently verifiable components. This technique is referred to as compositional reasoning. For example, a large superblock component can be partitioned (often quite naturally) into a set of smaller block and sub-block components. When verifying a property of one of these partitioned components, we must specify a set of constraints that model the behavior of the other components (that is, the environment for the component under verification).

We define a constraint as follows:

Definition 5: constraint: A condition (usually on the input signals) that limits the set of behavior to be considered by the formal verification engine. A constraint may represent real requirements on the environment in which the design is used, or it may represent artificial limitations imposed in order to partition the verification task.

Gradual semi-exhaustive verification. Although in theory, compositional reasoning using constraints sounds attractive, when applying formal property checking within an industrial setting, a more modest approach is generally used. We refer to this approach as *gradual semi-exhaustive formal verification via restrictions.* The advantage of this approach is that it has the potential of flushing out complex bugs as quickly as possible using formal verification to search a large state space.

Essentially, this approach is a gradual development of a formal verification environment around the RTL component that was selected using restrictions.

This approach has the following benefits:

- Allows us to control the state space explored to prevent state explosion using restrictions
- Enables us to initially turn off portions of the design's functionality - and then gradually turn on additional functionality as we validate the design under a set of restrictions
- Allows us to refine the constraint model into more general assumptions without initially encountering state explosion
- Provides an easier method of debugging by selecting, and thus controlling, the functionality in the environment that is enabled

We define a restriction as follows:

Definition 6: restriction: A statement that the design is constrained by a given artificial property and a directive to verification tools to consider only paths on which the given property holds.

A restriction may reduce a set of opcodes to a smaller set of legal values to be explored during the formal verification search process. Or a restriction may limit the component's mode settings to read only during one phase of a proof, and then re-prove with a write mode restriction. Other examples include restricting the upper eight bits of a 16-bit bus to a constant value while letting the lower eight bits remain unconstrained during the formal verification search, and then shifting the restriction to a new set of bits and re-proving with the new bus restrictions. It is important to note that even with the use of restrictions, the number of scenarios that the formal verification engine explores is very large, and complex errors will be detected under these conditions.

A variant of this technique has been used in industrial settings to prove the correctness of properties, and it is based on the notion of inductive proofs. First, the designer proves that the property holds for the reset state of the circuit. Then, we constrain the formal method of the design to *look like* it is the base case for then proving the inductive step. As a major advantage, this method can be very effective, as it optimally uses model checking for proving circuits exhaustively for a very short number of cycles.

One of the difficulties encountered is that usually verification engineers must have a very good understanding of the property that needs to be proven and the design itself, because if they underconstrain the design, they may get a false positive. On the other hand, if they overconstrain the design, they may get a false negative. Usually, the designer starts from a very weak constraint (to get a false positive) and incrementally strengthens it until the property becomes true. Although this may get fast proof results, we cannot tell in general if we have overconstrained the model.

Exhaustive proofs. The second technique used in an industrial setting, which is often the outcome of a gradual semi-exhaustive verification approach, is to relax the restrictions into general interface assumptions in an attempt to *prove* properties on the partitioned component. The advantage of performing the semi-exhaustive verification approach first using restrictions, as opposed to exhaustive proofs, is that if we cannot prove the property under the restriction, then we cannot prove it using general assumptions. Hence, we must employ other techniques (such as abstraction) if a proof is required.

We define an assumption as follows:

Definition 7: assumption: A statement that the design is constrained by a given property and a directive to verification tools to consider only paths on which the given property holds.

Note the subtle distinction between assumptions and restrictions related to our goal of applying formal verification technology in an industrial setting. For restrictions, our goal is to find bugs and clean up the partitioned components of the design using formal techniques. We are under no obligation to validate restrictions (either in simulation or formal verification). Using assumptions, however, our goal is to prove correctness, which can be a more difficult task. Often, we convert assumptions into assertions, which we then attempt to prove on neighboring components of the design. This strategy is known as *assume-guarantee reasoning* [8]. If an assumption is too difficult to formally prove, we use simulation to validate these assumptions as interface assertions.

5.2 Formal property checking role

In this section, we discuss the role formal property checking plays during various phases within a design flow. The first step in the process is to identify good property candidates that provide a clear return on investment (ROI) for the effort involved in the formal verification process and likelihood for success (LFS). Examples include properties related to portions of the design that:

- have historically resulted in respins due to bugs (hence, ROI)
- are estimated to be difficult to verify (or it will be difficult to achieve high coverage) using traditional simulation means (hence, ROI)
- are contained in control-intensive logic vs. data path logic (hence, LFS)
- are supported with enough bandwidth from the design team to adequately define required environment constraints when a full proof is required (hence, LFS)

Good property candidates for formal verification can occur at multiple levels of abstraction and phases of the design process. The level of expertise required for success at each phase varies depending on the verification goals.

Architectural verification. Formal verification has been successfully applied to proving architectural properties on shared memory consistency protocols (for example, cache coherence or sequential consistency protocols) as well as other architectural considerations (for example various arbitration schemes). The goal of this phase of formal verification is to flush out high-level architectural bugs prior to RTL implementation. However, successful architectural formal verification in an industry setting, in general, requires a verification team with a high level of expertise. In part, this expertise requirement comes from the need to create abstract models of the system that are *formal-friendly*.

Concurrent design and verification. Formal verification within an industry setting can be applied early during the RTL development phase in an attempt to flush out bugs prior to module integration into the system verification environment. In general, this is a low-effort task (which could be higher depending on the particular engineer's goals). As the engineer codes assertions into the RTL implementation, formal property checking combined with interface *restrictions* attempt to find bugs. If time permits, the verification engineer might attempt to extend this gradual semi-exhaustive approach into a full proof.

Block-level regression. Formal verification, when applied to the block level, offers much more than a low-effort, early bug-hunting tool. On the contrary, the strategy offers a means to deliver high quality blocks to the chip integration environment. Although the initial effort, before chip integration, does allow for early bug hunting, formal property checking's value extends beyond the initial stage. To provide a quick path for finding bugs and saving precious debug time during regression, it can also be performed every time the team modifies the block-level RTL code. This especially makes sense after a team makes the initial constraint investment at the block-level, which allows a formal tool to quickly prove the block-level assertions.

Post-silicon verification. We have successfully applied formal property checking during post-silicon verification. When a bug is identified in the lab, a formal test environment is created around the RTL implementation containing the bug. A property associated with the bug is created, and then the error is demonstrated on the RTL model using formal property checking combined with a formal testbench (that is, environmental properties used as constraints). Once the corrected RTL implementation is available, it is instantiated into the formal testbench and the formal property checker is used again to verify the fix.

6. Assertions and synthesis

Assertions can be synthesized for both automatic verification and for on-line assertion validation. A synthesizable assertion can provide an automatic model that serves as a crosscheck against formal verification. For on-line assertion validation, an assertion can be checked during the lifetime of a design. In this section, we show how assertions can be used in live designs to guarantee that the design works as expected.

6.1 On-line validation

On-line validation is a technique that checks for design correctness during circuit operation. It uses the idea of *white-box verification* to provide a live circuit with enough observability so that any incorrect behavior that has not been captured during the design simulation and prototyping phases can still be captured during the circuit operation.

As most designs evolve over time, this technique can be used to fix design bugs for the next generation of chips and circuit debugging and correction in FPGAs.

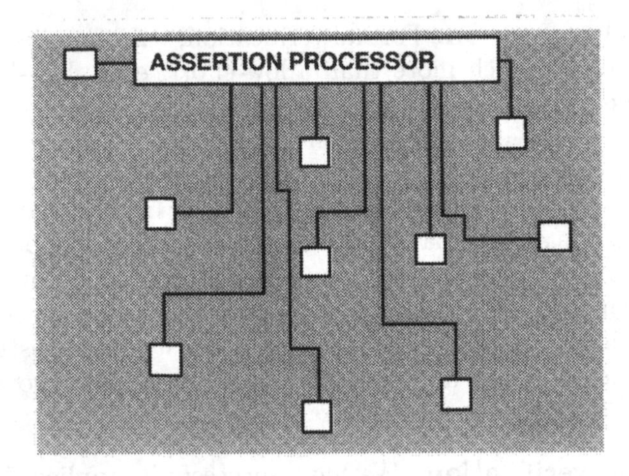

Figure 5.4. Diagram of assertion processor framework

Figure 5.4 is a diagram of a framework that must be added to an integrated circuit to process assertions. There are three main components to this framework: the assertions (square white boxes) that are distributed over the integrated circuit representing the conditions that must be checked during the circuit operation; the assertion processor, which is a circuit designed to process the results of the assertions and take proper action, being as simple as raising an error pin to put the system in halt mode or even to communicate with an embedded processor to dispatch an error correction routine; and a routing scheme that communicates the results of the assertions to an assertion processor.

6.2 Synthesizable assertions

Which assertions are suitable for on-line validation of integrated circuits? The reader should recall the three types of properties that were described in Section 2: *safety, liveness* and *fairness.*

Of these three, *safety* properties are the only type that can be used for on-line validation, because we can analyze the past history of a design to ascertain that nothing bad has happened to the circuit yet. Although running a live circuit for a long time does not guarantee that the property will ever hold false, we can ensure that it has not violated the property yet. The other types of properties, namely *liveness* and *fairness*, depend either on future events to ascertain validity or infinite traces.

A synthesizable assertion can be obtained if we code an assertion using a synthesizable RTL subset of a hardware description language, such as Verilog HDL or VHDL. Example 5.11 shows the code for an **assert_never** assertion, while Figure 5.5 illustrates the circuit.

Some researchers have proposed that the complexity of a synthesized assertion lies on the conditions passed to the assertion, and not in the assertion itself. We partially agree with this statement because the conditions that a designer wants to check are usually encoded in some expression inside the circuit description. By carefully choosing the correct form of the expression (for example, favoring re-usable expressions by the synthesis tool), a designer can avoid the regeneration of long and complex expressions that will add to the overall circuit size.

We leave to the reader to contrast the original **assert_never** from the description in the Open Verification Library with this simplified version to understand the conversion from a simulation assertion to a synthesizable assertion.

Assertions that can be synthesized can be classified as either deterministic or non-deterministic. Deterministic assertions trigger after a fixed number of cycles if a design failure is detected. An example of this type of assertion is **assert_never** , which triggers if a condition is true. On the other hand, non-deterministic assertions will be tested only after an initial (or triggering) event occurs, and as a result, may never be tested. Examples of these assertions are **assert_window** and **assert_time**. The reader should be aware of non-deterministic assertions, as they may never fire in a design not because the property being checked is true, but because the initial event never triggers.

Example 5.11 A property's verification layer expressed in PSL

```
module assert_never (ck, reset_n, expr);
  input ck, reset_n, expr;
  reg result;
  always @(posedge ck)
    if (reset_n == 1'b0)
      result <= 1'b1;
    else
      result <= (expr == 1'b0);
  initial result = 1'b1;
endmodule
```

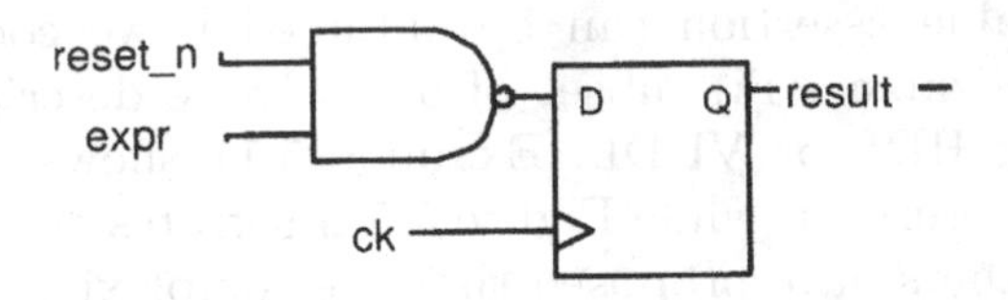

Figure 5.5. **assert_never** synthesized circuit

6.3 Routing scheme for assertion libraries

A synthesized assertion has to connect on one side to the integrated circuit itself and on the other side to the assertion processor. Note that a portion of the assertion library has to work on the speed of the design itself, but the portion connected to the assertion processor is usually subject to much less stringent time requirements.

One of the consequences of this fact is that routing from the assertion to the assertion processor does not have to be implemented in a one-to-one manner. We can leverage the work on the scan-chain (IEEE-1149 standard [IEEE-1149]) to implement a routing scheme that communicates the results of the assertions to an assertion processor.

In Figure 5.6-a, we present a typical assertion module from the Open Verification Library. In order to use it in a scan-chain architecture we defined extra pins, three inputs and two outputs, as shown in Figure 5.6-b. Table 5.1 contains descriptions of each additional signal.

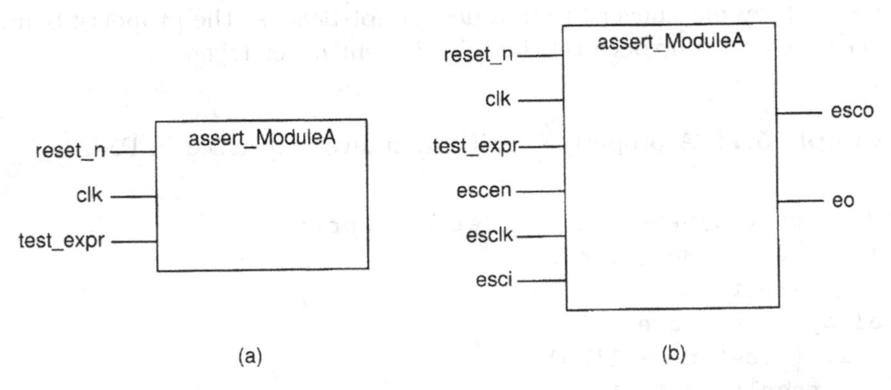

Figure 5.6. Extension to assertion library to enable distributed routing scheme

To better understand how this extended circuit works, we present in Figure 5.7 the routing scheme using scan-chain and in Figure 5.8 the

Table 5.1. Added pins for synthesizable assertion library extension.

Signal	*Description*	*I/O*
Escen	Error Scan Enable	Input
Esclk	Error clock	Input
Esci	Error Scan Input	Input
Esco	Error Scan Output	Output
Eo	Error Output	Output

synthesized circuit for `assert_never`. When an assertion fails, the pin `eo` will signal the assertion processor of a failed assertion. Immediately, the assertion processor will start scanning the assertion sequence (using `esck` clock) to determine the failed assertion. Note that in this case however, the assertion processor will not know which assertion failed immediately, requiring an additional scan sequence and adding complexity to the assertion processor. Figure 5.9 shows a possible timing sequence that can be issued by the assertion processor to determine which assertion has failed.

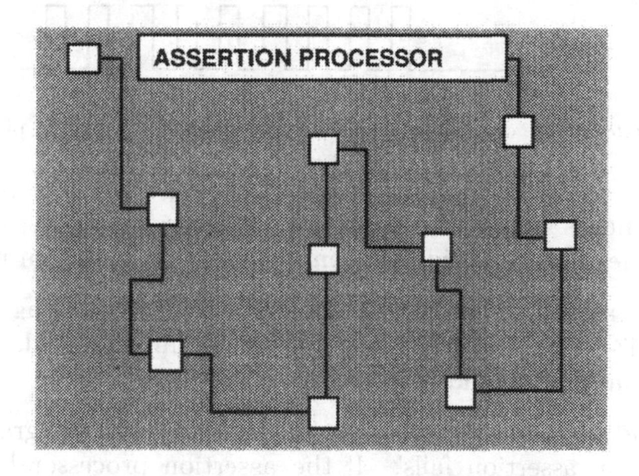

Figure 5.7. Scan-chained routing

6.4 Assertion processors

All of our discussion leads to the implementation of an assertion processor. The assertion processor can be as complex as a small micro-

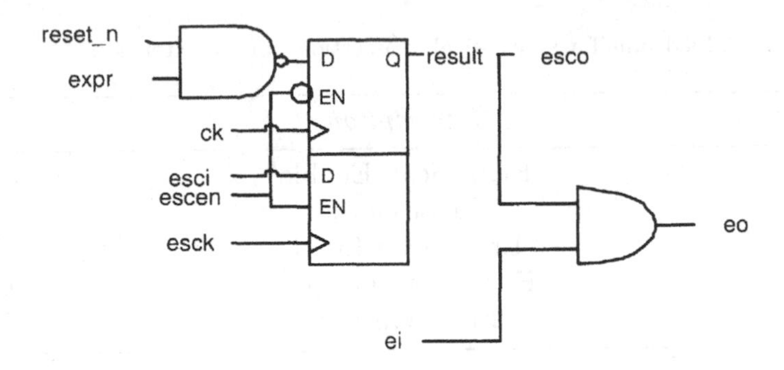

Figure 5.8. Synthesized circuit for modified **assert_never**

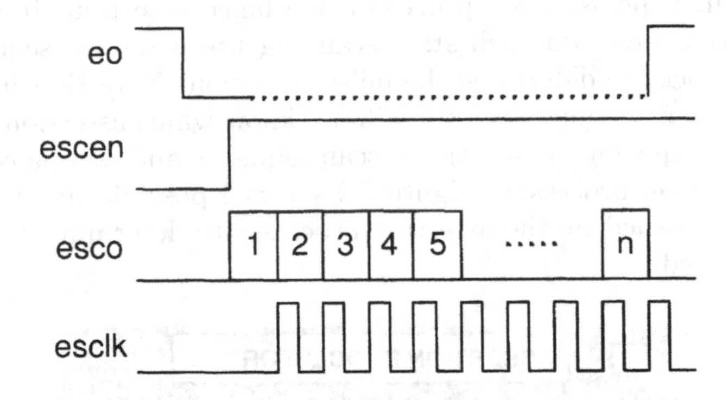

Figure 5.9. Time diagram of assertion scan issued by assertion processor

processor and as simple as a single pin. The designer must answer the following questions to decide the complexity of an assertion processor.

- Is the assertion processor responsible for determining what happened? If yes, a full scanner has to be implemented. Otherwise external pins should be provided.

- Should the assertion processor interact with the integrated circuit when an assertion fails? If the assertion processor has to stop processing inside the chip, or if it has to interrupt an embedded processor, additional complexity has to be added to the assertion processor.

A simpler assertion processor enables an entire system to be managed by an external assertion processor. On the other hand, a complex assertion processor can deal with more difficult situations, such as the example mentioned above.

Note that the assertion mechanism described earlier allows us to determine which assertion has failed but not the conditions under which the assertion failed. As a result, a more complex assertion processor has to be implemented if we need to know the conditions under which the circuit has failed. In this case, the assertion processor should have access to a local storage and it should be able to stop the processor from continuing.

6.5 Impact of Assertions in Real Circuits

In [12], several assertions from the Open Verification Library were synthesized for a Xilinx FPGA, with bit ranges and delays ranging from 1 to 32. The sizes of the assertions ranged from 3 LUTs and 1 flip-flop (in the case of `assert_always`) to 149 LUTs and 67 flip-flops (in the case of `assert_change`).

It has been suggested that just synthesizing the assertions does not give a good measure because RTL languages enable designers to write very complex expressions when instantiating assertions, and these expressions would not be accounted for in the synthesis results. However, in real designs, the expressions specified in the assertion instances generally correspond to some cone of logic previously specified by the designer. As a result, its impact should be minimal.

7. PCI property specification example

In this section, our goal is to demonstrate a process of translating a set of natural language requirements into a set of properties. We have chosen examples from the Peripheral Component Interconnect (PCI) specification [1]. Please note that it is not our intention to fully specify all functional requirements of the PCI – we leave this as an exercise for the reader.

You will note that many of the properties we specify in this section are at a transaction level. Protocol specification and verification at a transaction level is more efficient than at a signal interaction level. Transaction level specification not only permits more efficient test stimulus generation, it also enables debugging and measuring functional coverage at a higher level of abstraction. Nonetheless, specifying transaction level properties is not always efficient for formal verification, and partitioning the transaction level property into a set of simpler properties can yield better results.

Transactions are conveniently constructed by partitioning the behavior definition into a set of sequence specifications, with each sequence representing a specific behavior segment of a transaction. These sequences are then combined to form a more complex bus transaction specification. We recommend that interface protocol or transaction specification occur prior to coding the RTL.

7.1 PCI overview

The PCI local bus is an industry standard, high performance 32- or 64-bit local bus architecture with multiplexed address and data lines. The bus was defined with the primary goal of establishing an industry standard, high performance, low cost interconnect mechanism between highly integrated peripheral controller components, peripheral add-in boards, and processor/memory systems.

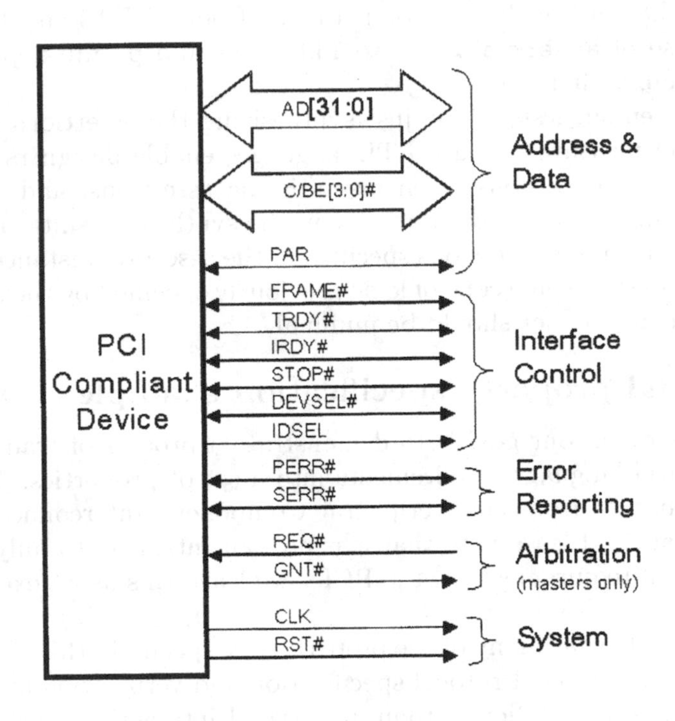

Figure 5.10. PCI interface

We begin our discussion of creating a PCI formal specification by illustrating the bus interface required pin list as shown in figure1-8. This is followed by a brief description for each required PCI signal. Finally, we demonstrate how to convert a natural language specification of the PCI bus protocol into a set of assertions.

A PCI bus transaction consists of an *address phase* followed by one or more *data phases*. During the address phase, the `C/BE[3:0]#` bus command indicates the type of transaction. During the data phase, C/BE[3:0]# are used as Byte Enables.

Note that the # symbol at the end of the signal name indicates an active low signal. For our examples, we convert the # symbol into "n" as part of the name to indicate an active low signal.

7.2 PCI master reset requirement

In this section, we demonstrate how to translate a simple PCI reset requirement, stated in Section 2.2.1 of the PCI Local Bus Specification [1]. The PCI reset requirement is stated as follows:

To prevent AD, C/BE#, *and* PAR *signals from floating during reset, the central resource may drive the* RST# *lines during reset (bus parking) but only to a logic low level; they may not be driven high.*

In Example 5.12, we have written a PSL assertion to check that the AD, C/BE#, and PAR signals are never driven high during reset.

Example 5.12 PSL assertion for PCI signal reset requirements.

```
default clock = (posedge clk);
assert always (rst_n==0) -> !(|{ad, cbe_n, par});
```

Note that we have used the Verilog *reduction* or operator to determine if any bit in this example is a logical one. The same assertion could be specified using a Verilog OVL implication monitor as shown in Example 5.13.

Example 5.13 OVL assertion for PCI signal reset requirements.

```
assert_always mstr_reset (clk, !rst_n, !(|{ad,cbe_n,par});
```

7.3 PCI burst order encoding requirement

The memory address space for the PCI is defined by the bits AD[31:2]. The lower two bits (that is, AD[1:0]) are encoded to indicate the order in which the master is requesting the data transfer, as defined in Section 3.2.2.2 of the PCI Local Bus Specification. Table 5.2 specifies the legal burst order encoding for memory transactions. Hence, address bit AD[0] must never be set to an active high value for a memory transaction burst order request.

Table 5.2. Burst order encoding.

AD[1	*AD[0]*	*burst order*
0	0	linear increment
0	1	reserved
1	0	cache wrap mode
1	1	reserved

In Example 5.14, we code a PSL assertion to validate a correct memory burst order request. Note that this assertion uses a sequence to define a memory address phase sequence (that is, a falling edge of FRAME#, along with the decoding of a memory transaction from the bus command C/BE#). Whenever this prefix sequence occurs, then bit AD[0] must always be active low for a valid burst order encoding.

The PCI memory address phase is described by defining the sequence SERE_MEM_ADDR _PHASE, which matches sequences containing a falling edge of FRAME# combined with a decoding of a memory command. This forms a prefix sequence, which implies that the reserved AD[0] is not active high.

Example 5.14 PSL burst order encoding assertion.

```
'define mem_cmd ((cbe_n == .MEM_READ) || \
                 (cbe_n == 'MEM_WRITE) || \
                 (cbe_n == 'MEM_RD_MULTIP) || \
                 (cbd_n == 'MEM_RD_LINE) || \
                 (cbd_n == 'MEM_WR_AND_INV))


sequence SERE_MEM_ADDR_PHASE={frame_n;!frame_n && mem_cmd};
property PCI_VALID_MEM_BURST_ENCODING =
   always {SERE_MEM_ADDR_PHASE} |-> {!ad[0]}
      abort !rst_n @(posedge clk);


assert PCI_VALID_MEM_BURST_ENCODING;
```

7.4 PCI basic read transaction

In this section, we demonstrate (via a simplified example) another transaction-level property, which we construct by partitioning the transaction into a set of partial behaviors specified as sequences. A PCI basic read operation consists of the following phases:

- an *address phase*, which for a basic read consists of a single address transfer in one clock
- a *data phase*, which includes one transfer state plus zero or more wait states

The address phase occurs on the first clock cycle in which `FRAME#` is asserted. For a basic read transaction, there must be at least one turn around cycle between the address phase and the data phase. A data phase completes when an active `IRDY#` and either an active `TRDY#` or `STOP#` is clocked. The read transaction completes when `FRAME#` becomes inactive. In reality, there are numerous transaction-terminating conditions defined in section 3.3.3 of the PCI specification that can be initiated by either the master or target (for example, *timeout, abort, retry, disconnect*). For our PCI basic read operation, our goal is to demonstrate how to build a transaction through a set of sequence specifications. Hence, we have chosen to simplify our example and ignore these special terminating cases. We leave it to the reader to modify our example by specifying all terminating conditions.

Byte enable requirement. Section 3.3.1 (page 47) of the PCI Local Bus Specification states the following requirement associated with a read transaction:

The `C/BE#` *output buffers must remain enabled (for both read and writes) for the first clock of the data phase through the end of the transaction*

Example 5.15 demonstrates how to specify a PCI basic read transaction as specified with the `C/BE#` output buffer requirement. The property PCI_READ_TRANSACTION begins with an address phase (that is, SERE_RD_ADDR_PHASE). We then specify a sequence that describes the initial required turn around cycle (that is, SERE_TURN_AROUND), which occurs the first clock after the address phase. Then, the `C/BE#` signals remain unchanged throughout the remaining data phase `cbe_n==prev(cbe_n)`, that is, throughout SERE_DATA_PHASE.

When specifying protocol requirements, we have the choice of creating a complex property that captures all requirements required for the transaction, or partitioning the different requirements of the transaction into a set of simpler properties. For example, for simplicity we decided not to specify the read transaction latency requirements in Example 5.15 for either the bus target or master (as defined in Section 3.5 of the PCI specification). Hence, we could either modify our assertion example by directly writing in the additional bus latency requirements or we could create a separate simpler property for the latency requirements.

Example 5.15 PSL basic read transaction assertion.

```
'define data_complete
                    ((!trdy_n||!stop_n)&&!irdy_n&&!devsel_n)
'define end_of_transaction (data_complete && frame_n)
'define adr_turn_around (trdy_n & !irdy_n)
'define data_tranfer
                    (!trdy_n&&!irdy_n&&!devsel_n&&!frame_n)
'define wait_state ((trdy_n || irdy_n) && !devsel_n)
'define cbe_stable (cbe_n==prev(cbe_n))

'define read_cmd ((cbe_n == 'IO_READ) || \
                    (cbe_n == 'MEM_READ) || \
                    (cbe_n == 'CONFIG_RD) || \
                    (cbe_n == 'MEM_RD_MULTIP) || \
                    (cbe_n == 'MEM_RD_LINE))

sequence SERE_RD_ADDR_PHASE={frame_n;!frame_n && read_cmd};
sequence SERE_TURN_AROUND = {adr_turn_around};
sequence SERE_DATA_TRANSFER =
{
    {{wait_state[*];data_transfer}[1:inf]}
};
sequence SERE_END_OF_TRANSFER = {data_complete && frame_n};
sequence SERE_DATA_PHASE =
{
    {{SERE_DATA_TRANSFER};{SERE_END_OF_TRANSFER}} &&
     {cbe_stable}
};

property PCI_READ_TRANSACTION =
    always {SERE_RD_ADDR_PHASE} |=>
       {SERE_TURN_AROUND; SERE_DATA_PHASE} abort
           !rst_n @(posedge clk);

assert PCI_READ_TRANSACTION;
```

8. Summary

In this chapter, we discussed the important role that property specification plays in an assertion-based verification flow. In the future, we predict that design and verification will become property-based. Through the standardization of property languages, such as PSL, we foresee new and exciting EDA markets emerging, opening the door for improved productivity, and all made possible through assertion-based design practices. Once specified, properties may be used across multiple design and verification technologies, such as simulation, formal verification, and synthesis.

References

[1] *PCI Local Bus Specification.* PCI Special Interest Group, revision 2.2 edition, 1998.

[2] *Accellera proposed standard Property Specification Language (PSL) 1.0*, 2003.

[3] Abarbanel, Y., Beer, I., Gluhovsky, L., Keidar, S., and Wolfsthal, Y. Focs–automatic generation of simulation checkers from formal specifications. In *Proceedings of Computer Aided Verification*, 12th International Conference, pages 414–427, 2000.

[4] Clarke, E. and Emerson, E. A. Design and synthesis of synchronization skeletons using branching time temporal logic. LNCS 407. Logic of Programs: Workshop, 1981.

[5] Clarke, E., Grumberg, O., and Peled, D. *Model Checking.* The MIT Press, 2000.

[6] Foster, Krolnik, A., and Lacey, D. *Assertion-Based Design.* Kluwer Academic Publishers, 2003.

[7] Foster, H. and Coelho, C. Assertions targeting a diverse set of verification tools. Proc. Intn'l HDL Conference, 2001.

[8] Grumberg, O. and Long, D. Model checking and modular verification. In *ACM Transaction on Programming Languages and Systems*, pages 843–872, 1994.

[9] Kantrowitz, M. and Noack, L. I'm done simulating; now what? verification coverage analysis and correctness checking of the decchip 21164 alpha microprocessor. In *Proc. Design Automation Conference*, pages 325–330, 1996.

[10] Kripke, S. Semantic considerations on model logic. In *Proceedings of a Colloquium: Modal and Many valued Logics*, volume 16, pages 83–94, 1963.

[11] Kropf, T. *Introduction to Formal Hardware Verification.* Springer, 1998.

[12] Nacif, J., de Paula, F., Foster, H., Jr., C. C., Sica, F., Jr., D. S., and Fernandes, A. An assertion library for on-chip white-box verification at run-time. In *Proc. Latin American Test Workshop*, 2003.

[13] Pnueli, A. The temporal logic of programs. In *18th IEEE Symposium on foundation of Computer Science.* IEEE Computer Society Press, 1977.

[14] Taylor, S., Quinn, M., Brown, D., Dohm, N., Hildebrandt, S., Huggins, J., J., and Ramey, C. Functional verification of a multiple-issue out-of-order, superscalar alpha processor-the dec alpha 21264 microprocessor. In *Proc. Design Automation Conference*, pages 638–643, 1998.

Chapter 6

FORMAL VERIFICATION FOR NONLINEAR ANALOG SYSTEMS: APPROACHES TO MODEL AND EQUIVALENCE CHECKING

Walter Hartong

Cadence Design Systems, Feldkirchen, Germany

hartong@cadence.com

Ralf Klausen

Institute of Microelectronic Systems, University of Hannover, Germany

klausen@ims.uni-hannover.de

Lars Hedrich

Institute of Microelectronic Systems, University of Hannover, Germany

hedrich@ims.uni-hannover.de

Abstract In this contribution, we present equivalence and model checking methods for nonlinear analog systems. Both approaches are based one the system's nonlinear state space description. The equivalence checker computes a nonlinear transformation of the state space descriptions into a canonical form. Thus, the input/output behavior of the specifying and the target system can be compared independently of the different state representations. The model checking approach uses an automatic state space subdivision method to transfer the continuous state space into a discrete model retaining the essential analog dynamics. The analog system properties are described in an extended CTL language. Experimental results show the feasibility of both approaches.

Keywords: Analog, formal verification, model checking, equivalence checking

R. Drechsler (ed.), Advanced Formal Verification, 205-245.

1. Introduction

Growing design size and shorter design cycles increase the need for powerful validation methods for digital as well as for analog and mixed signal systems. Formal verification methods for analog systems can complement the existing methods for digital systems and hybrid systems. They can be used for verifying the behavior of digital library cells on transistor level in order to close the chain of evidence to digital formal verification approaches or directly for verifying the analog behavior of an analog cell.

The goals used in digital formal verification algorithms are also valid for analog verification. Two major method classes (equivalence checking and model checking) can be defined for analog systems as follows.

Definition 6.1 *Equivalence checking for analog systems proves or disproves that the input/output behavior of a target system is equal to that of a specifying system. The proof is valid for all possible input stimuli and for the entire system behavior.*

Definition 6.2 *Model checking for analog systems proves or disproves given properties of a target system. The proof is valid for all possible input stimuli and for the entire system behavior.*

However, some distinctions have to be made. Since analog domains are continuous in time and value, calculating exact equivalences of two values is difficult. Therefore, computing formal statements means to deal with tolerances, intervals, regions, or approximations.

Furthermore, the system description given by differential equations is not solvable analytically in general. Hence, an approximation of the solution in a numerical process is necessary. Due to this approximation, a theoretical proof, as required for a formal verification algorithm in a narrower sense, can not be achieved for nonlinear analog systems. However, this is true for any formal verification process on physical systems since the achievable accuracy is limited by the system model used [24, 29].

In this chapter, two approaches to formal verification are presented. The first one implements an equivalence checking method comparing two analog system descriptions. The second one is a model checking algorithm, enabling the check of certain specification properties versus the target system.

2. System Description

In contrast to digital logic, analog circuits are continuous in signal values and time. Checking the exact equivalence or exact properties of an analog system is not reasonable, because even a small deviation of a system parameter may lead to a negative verification result. Due to this, formal verification of analog systems has to handle deviations of the system's behavior. In this chapter, two possible methods to deal with deviations are described. First, a deviation measure can be defined to indicate the error magnitude. If this error is below a predefined threshold, the systems are equal from a practical point of view.

Second, the system can be discretized by building boxes (intervals). These boxes can be handled like distinct sets of an infinite number of signal values enabling set based comparison. The comparison is exact due to set operations. Hence, the continuous character of analog signals disappears leaving some discretization errors.

Analog circuits can be modeled by circuit descriptions like netlists with corresponding device and behavioral models. The basic system description for both problems, equivalence and model checking, are non-autonomous systems of nonlinear first-order differential equations. The system of differential equations can be built by using the Modified Nodal Approach [23]. This approach uses Kirchhoff's node equations and additional device equations for special devices like voltage sources and inductors.

In this chapter, we restrict ourselves to single-input single-output (SISO) systems. However, the approaches described below can be easily extended to handle MIMO (multiple-input multiple-output) systems.

An analog system (Equation (6.1)) consists of n nonlinear first order differential algebraic equations (DAE system)

$$\mathbf{f}(\mathbf{x}(t), \dot{\mathbf{x}}(t), u(t)) = \mathbf{0} \Leftrightarrow \left\{ \begin{array}{l} f_{(1)}(\mathbf{x}(t), \dot{\mathbf{x}}(t), u(t)) = 0, \\ f_{(2)}(\mathbf{x}(t), \dot{\mathbf{x}}(t), u(t)) = 0, \\ \dots \\ f_{(n)}(\mathbf{x}(t), \dot{\mathbf{x}}(t), u(t)) = 0 \end{array} \right\} \tag{6.1}$$

with the input $u(t)$, the time t, the vector $\mathbf{x}$ (Equation (6.2)) of n system variables (e.g. node voltages and currents)

$$\mathbf{x}(t) = \left[x_{(1)}(t), x_{(2)}(t), ..., x_{(n)}(t)\right]^T, \tag{6.2}$$

and its time derivative $\dot{\mathbf{x}}(t)$.

For verification tasks, an output variable $y(t)$ has to be identified for each system. In general, this is done by defining an output equation

$$g(y(t), \mathbf{x}(t)) = 0\,. \tag{6.3}$$

In general, the output variable is a system variable $\mathbf{x}(t)$. In this case, g is a simple selection function in terms of $\mathbf{x}(t)$ and $y(t)$. The input $u(t)$

is not part of g, because direct influences from inputs to outputs can be always described by introducing additional variables and equations.

2.1 Analog Circuit Classes

Since analog circuits can be divided into different classes, the formal verification problem can be addressed by different methods. Three main classes of analog circuits can be distinguished based on the following criteria [20]:

- linear/nonlinear,
- static/dynamic,
- nominal values/parameter tolerances.

Our approaches deal with nonlinear dynamic circuits without tolerances. To our knowledge, there exists no formal verification approach for the most general circuit class, namely nonlinear dynamic circuits with parameter tolerances, due to time complexity problems. However, it is theoretically possible to extend the proposed algorithms in order to deal with tolerances.

2.2 State Space Description

State space descriptions are a widely used concept in the analog domain, especially in control theory, and of course in the digital domain. For all cases, the circuit behavior can be described by means of the state space enabling a canonical description and simplifying algorithmic access.

An electrical circuit can be described combining Equation (6.1) and Equation (6.3) to

$$\begin{aligned} \mathbf{f}(\mathbf{x}(t), \dot{\mathbf{x}}(t), u(t)) &= \mathbf{0}, \\ g(y(t), \mathbf{x}(t)) &= 0. \end{aligned} \tag{6.4}$$

A subset of system variables $\mathbf{x}(t)$ is called the state variables $\mathbf{x}^{\mathsf{z}}(t)$:

Definition 6.3 *A state variable $x_i^{\mathsf{z}}(t)$ is a variable, whose value can be chosen independently (at least valid under some constraints) for the solution of the DAE system (Equation (6.4)).*

The remaining variables are called non-state or algebraic variables $\mathbf{x}^{\mathsf{a}}(t)$. Both variable types build up the n system variables

$$\mathbf{x}(t) = [\mathbf{x}^{\mathsf{z}}(t), \mathbf{x}^{\mathsf{a}}(t)] \; . \tag{6.5}$$

The z state variables $\mathbf{x}^{\mathbf{z}}(t)$ span the state space. Every point in the state space represents a state of the system. If the input $u(t)$ is added as an additional dimension to the state space, the resulting space is called extended state space.

A SISO system, described by the equation system $\mathbf{f}$ and the output equation g, can be graphically represented by a vector field of the derivatives of the state variables $\dot{\mathbf{x}}^{\mathbf{z}}(t)$ in the extended state space and a scalar field of the output variable $y(t)$ (see Figure 6.1). The system's dynamics are described by the vector field of the derivatives of the state variables. Trajectories of the system starting from a state in the state space can be constructed simply by following the time derivatives vectors. These trajectories are computed in transient analysis of circuit simulation for example. Both fields together describe the entire nonlinear dynamic input/output behavior of the system.

In general, the state variables are those variables for which initial values can be independently chosen regarding the corresponding initial value problem of the DAE system (see also Section 2.2.2). A state variable occurs at least once in time derivative form $\dot{x}_i^{\mathbf{z}}(t)$ within the DAE system. In some special cases constraints might exist, which limit the free choice of the state variables, for example a linear dependent combination of two state variables' derivatives result in only one freely selectable state variable. These cases will be neglected in the further explanation. Moreover, hidden constraints from equations result in even more complex dependencies, measured by the index of the equation system.

2.2.1 Index.

Systems of ordinary differential equations (ODEs) have no algebraic variables. If algebraic variables occur, the system is called a system of differential algebraic equations (DAEs) [31]. A widely used measure for the solveability of DAE systems is the index. Many index definitions exist [15]. In the case of nonlinear equations, the differentiation index is used [14]. For linear or linearized equations, the Kronecker index [27] can be used. An ODE system has the index 0, a DAE system has at least the index 1. For systems with index 2 or higher, indicating that there are hidden constraints in the equations, the solveability of the system is not straightforward and additional techniques have to be used. Some techniques are described in [11]. Our verification approaches deal with index 0 and 1 problems, but can be extended by appropriate techniques to higher indices.

2.2.2 Solving a DAE System.

The basic task in analog system analysis is to solve the nonlinear differential equation system

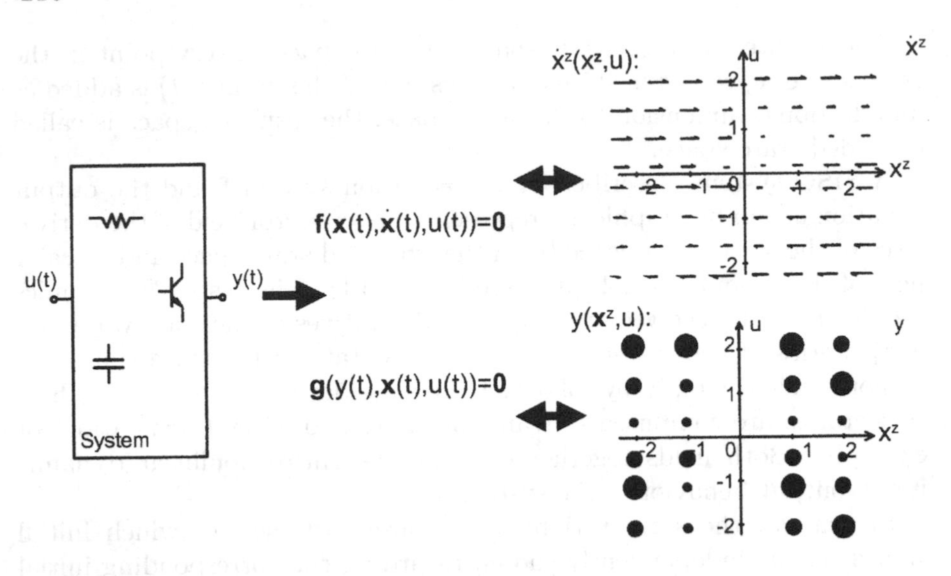

Figure 6.1. Equivalent descriptions of a system as implicit state space equations and as vector and scalar fields in the extended state space.

(Equation (6.4)). Assuming given initial values $\mathbf{x}^{\mathbf{z}}(t_0)$, the problem to be solved is the initial value problem (IVP).

In the index 0 case with no algebraic variables $x^{\mathbf{a}}(t)$, the DAE-System can be solved theoretically for $\dot{\mathbf{x}}(t) = \mathbf{f}'(\mathbf{x}(t), u(t))$. Accordingly the solution of the initial value problem can be calculated by directly integrating the DAE system

$$\mathbf{x}(t_0 + \Delta t) = \mathbf{x}(t_0) + \int_{t_0}^{t_0+\Delta t} \mathbf{f}'\left(\mathbf{x}(\tau), u(\tau)\right)\, d\tau\,. \tag{6.6}$$

The integration can be done in an analytical or - in most cases - in a numerical way. $y(t)$ can be computed easily by evaluating g for $\mathbf{x}(t_0 + \Delta t)$.

However, electrical circuits are usually not index 0 systems. Additionally, the equations $\mathbf{f}$ can not be solved for $\dot{\mathbf{x}}(t)$. Thus, we have to integrate an implicit DAE system numerically. The initial value $\mathbf{x}^{\mathbf{z}}(t_0)$ for each state variable $x^{\mathbf{z}}$ has also to be given. One of the most important numerical integration methods is the backward Euler formula. In this case, the time derivatives $\dot{\mathbf{x}}^{\mathbf{z}}$ are replaced by the numerical integrator in Equation (6.4)

$$\mathbf{x}(t_n), y(t_n) = \left\{ \mathbf{x}(t_n), y(t_n) \;\middle|\; \begin{array}{rcl} \mathbf{f}\left(\frac{\mathbf{x}^z(t_n)-\mathbf{x}^z(t_{n-1})}{\Delta t_n}, \mathbf{x}(t_n), u(t_n)\right) & = & \mathbf{0} \\ g(y(t_n), \mathbf{x}(t_n)) & = & 0 \end{array} \right\} . \quad (6.7)$$

The solution $\mathbf{x}(t_n), y(t_n)$ is computed by a numerical solver for the resulting algebraic equation system, for example by Newton-Raphson iteration.

In circuit simulations, namely transient analysis, the DC solution of the system is chosen as $\mathbf{x}(t_0)$. The following solutions are sequentially computed using the newly calculated value $\mathbf{x}(t_n)$ as an initial value for the next time step.

2.2.3 Linearized System Description. The previous paragraph dealed with nonlinear large signal values of the analog system. It is often an advantage to have a linear or linearized system description enabling the transformation to the frequency domain and the use of the powerful theory for linear systems. As we will see later, a linearized system description is also used for the equivalence checking approach (see Section 3.3).

For small deviations around a solution $\mathbf{x}$, a locally linearized system of equations resulting from $\mathbf{f}$ can be set up

$$\begin{aligned} \mathbf{C} \cdot \dot{\mathbf{x}}(t) + \mathbf{G} \cdot \mathbf{x}(t) &= \mathbf{b} \cdot u(t)\,, \\ y &= \mathbf{r}^T \cdot \mathbf{x}(t)\,, \end{aligned} \quad (6.8)$$

with the capacitance and admittance matrices

$$\mathbf{C} = \frac{\partial \mathbf{f}(\dot{\mathbf{x}}, \mathbf{x}, u)}{\partial \dot{\mathbf{x}}}, \mathbf{G} = \frac{\partial \mathbf{f}(\dot{\mathbf{x}}, \mathbf{x}, u)}{\partial \mathbf{x}}, \quad (6.9)$$

and the input and output vectors

$$\mathbf{b} = -\frac{\partial \mathbf{f}(\dot{\mathbf{x}}, \mathbf{x}, u)}{\partial u}, \mathbf{r}^T = -\frac{\partial g(\mathbf{x})}{\partial \mathbf{x}}. \quad (6.10)$$

3. Equivalence Checking

There exist some approaches to equivalence checking for hybrid systems which use known verification methodologies for digital systems verification and adapt them to analog or partially analog systems [22, 32]. They are restricted by verifying linear or piecewise linear analog systems. In [20, 18] some approaches to equivalence checking for circuits with tolerances are described. One approach deals with the non-general class of linear systems, another one with static, nonlinear systems.

On the other hand, there are a lot of simulation based techniques computing performance characteristics from a target circuit and comparing them to given specifications, like *worst-case analysis* [9, 2]. These approaches are based on probabilistic methods giving good and reliable results for the mentioned tasks. However, they are neither able to prove that a circuit with tolerance parameters fulfills a certain specification nor doing that for all possible input stimuli, which is demanded for formal verification.

The following approach deals with the class of nonlinear dynamic circuits and is restricted to nominal parameters. Therefore, it is not possible to perform an inclusion proof as in the approaches using parameter tolerances [19]. Accordingly the target systems and specifications need not to be distinguished.

3.1 Basic Concepts

The basic idea of the algorithm for nonlinear dynamic system verification is to compare two geometrical descriptions of two systems A and B. These geometrical descriptions can be obtained by transforming the two equation systems into state space descriptions, as explained in Section 2.2. The objective is to determine whether the vector fields $\dot{\mathbf{x}}^{\mathbf{z}}_A(\mathbf{x}^{\mathbf{z}}_A, u)$, $\dot{\mathbf{x}}^{\mathbf{z}}_B(\mathbf{x}^{\mathbf{z}}_B, u)$ and the scalar fields $y_A(\mathbf{x}^{\mathbf{z}}_A, u)$, $y_B(\mathbf{x}^{\mathbf{z}}_A, u)$ are equal or not. Possible different behaviors of the systems result in differences in the scalar and vector fields and are evaluated using an appropriate error calculation.

3.1.1 Nonlinear Mapping of State Space Descriptions.

In general, two systems do not have the same internal state variables because they are represented by different implementations on possibly different levels of abstraction. Therefore, the simple method of computing the vector fields in state spaces directly is not able to identify systems with a similar input/output behavior.

For example, consider the two systems in Equation (6.11). The systems A and B are equal with respect to their input/output behavior.

$$\begin{array}{ll} \text{System A}: & \text{System B}: \\ f_A : \dot{x}_A + 3(x_A - x_A^{\frac{2}{3}} \cdot u) = 0 & f_B : \dot{x}_B + x_B - u = 0 \\ g_A : y_A - x_A = 0 & g_B : y_B - x_B^3 = 0 \end{array} \qquad (6.11)$$

f_B can be derived from f_A using the translation rule $x_A = x_B^3$. However, the differential equations and the vector and scalar fields are different (see Figure 6.3). Therefore, mapping functions $z_A = t_A(x_A)$ and

$z_B = t_B(x_B)$ have to be found which uniquely map the state variables x_A, x_B onto the so called virtual state variables z_A, z_B. The idea behind this step is to map the state space descriptions $\dot{\mathbf{x}}^{\mathbf{z}}_A(\mathbf{x}^{\mathbf{z}}_A, u)$, $\dot{\mathbf{x}}^{\mathbf{z}}_B(\mathbf{x}^{\mathbf{z}}_B, u)$ into a canonical representation $\dot{\mathbf{z}}_A(\mathbf{z}_A, u)$, $\dot{\mathbf{z}}_B(\mathbf{z}_B, u)$. The canonical representation should be a representation with decoupled, ordered states. In the linear case, this could be achieved with the Kronecker Normal Form (see Section 3.3.1). In the nonlinear case, the mapping has to be iteratively computed as we will see later.

As a result from the mapping into the canonical form, the vector field $\dot{z}_A$ in the virtual extended state space of the transformed system A can be compared to the vector field $\dot{z}_B$. For this example using $z_A = x_A^{\frac{1}{3}}$ and $z_B = x_B$ leads to two identical vector fields $\dot{z}_A$ and $\dot{z}_B$ (see Figure 6.3). The scalar fields y_A and y_B are transformed by the mapping functions accordingly.

Using this result, we can define our verification approach more precisely: Two systems with equal virtual state space vector fields $\dot{\mathbf{z}}_A$, $\dot{\mathbf{z}}_B$ and scalar fields y_A, y_B, resulting from mapping functions $\mathbf{t}_A$ and $\mathbf{t}_B$, have equal input/output behavior if

$$\begin{aligned} \dot{\mathbf{z}}_A(\mathbf{z}_A, u) &= \dot{\mathbf{z}}_B(\mathbf{z}_B, u) \;\wedge \\ y_A(\mathbf{z}_A, u) &= y_B(\mathbf{z}_B, u) \end{aligned} \tag{6.12}$$

is valid.

Since it is impossible, to compute the time derivatives of a DAE system (Equation (6.1)) analytically, a direct computation of the nonlinear mapping functions $\mathbf{t}_A$, $\mathbf{t}_B$ is also not possible. Therefore, a numerical computation including error calculation of Equation (6.12) has to be used.

3.2 Equivalence Checking Algorithm

Since the analytical generation of the mapping functions $\mathbf{t}_A$ and $\mathbf{t}_B$ is impossible, an iterative method is used. The idea is to linearize the system at particular sampling points and to use linear mapping matrices $\mathbf{F}_{r_A}$ and $\mathbf{F}_{r_B}$ to apply a local linear mapping. To consider the nonlinear behavior of the systems, the linearization has to be done in the whole state space. This leads to an algorithm, which samples the extended state space and compares the vector and scalar fields at every discrete sampling point.

3.2.1 Sampling the State Space. As shown above, the extended state space of the two systems is sampled in order to find the mapping functions iteratively (see similar sampling method in system identification [5]). The boundaries of the extended state space are determined by the maximum excitation of the variables and the input, which are defined by the user. Assuming a finite step size leads to a finite set of points. The step size of the sampling algorithm is determined by a step size control algorithm, described in [20]. The basic verification algorithm is shown in Figure 6.2.

Verification of nonlinear dynamic systems() {
  read netlists and setup nonlinear equations for both systems
  for every input value in a predefined range **do** {
    use DC analysis to obtain initial state vectors $\mathbf{x}_{0_A}$, $\mathbf{x}_{0_B}$
    compute linear mapping matrices $\mathbf{F}_{r_A}$, $\mathbf{F}_{r_B}$
    for every sample point in predefined ranges **do** {
      compute new step size $\Delta\mathbf{z}$
      compute state vectors for next sample point:
        $\mathbf{x}_{giv_A} = \mathbf{x}_{old_A} + \mathbf{F}_{r_A} \cdot \Delta\mathbf{z}$, $\mathbf{x}_{giv_B} = \mathbf{x}_{old_B} + \mathbf{F}_{r_B} \cdot \Delta\mathbf{z}$
      compute consistent sample points $\mathbf{x}_{cons_A}$, $\mathbf{x}_{cons_B}$
        using $\mathbf{x}_{giv_A}$, $\mathbf{x}_{giv_B}$
      compute linear mapping matrices $\mathbf{F}_{r_A}$, $\mathbf{F}_{r_B}$
      calculate $\varepsilon_{\dot{z}}$ (error of the state derivatives vector fields)
      calculate ε_y (error of the output scalar field)
      adjust $\mathbf{x}_{cons_A}$
      $\mathbf{x}_{old_A} = \mathbf{x}_{cons_A}$, $\mathbf{x}_{old_B} = \mathbf{x}_{cons_B}$
    }
  }
  check if $\varepsilon_{\dot{z}}$ and ε_y < errormargin for all sample points
}

Figure 6.2. Basic verification algorithm.

First, the generation of the differential equation systems described in Section 2 for both systems is necessary. Two loops are needed to sample the extended state space. In the outer loop, the input variable is increased stepwise from the given start to the end value. Inside this loop a DC analysis is accomplished to determine the operating point, used as initial state vector $\mathbf{x}_0$. Instead of computing the mapping functions $\mathbf{t}_A$ and $\mathbf{t}_B$, two local mapping matrices $\mathbf{F}_{r_A}$ and $\mathbf{F}_{r_B}$ are calculated, using a linearized state space description. This is necessary, because only for linear systems a direct calculation of the canonical representation and the corresponding mapping matrices can be found. A detailed description how to calculate the local linear mapping matrices is given in Section 3.3

The inner loop starts with the computation of the state vectors for the next sample point, using $\Delta\mathbf{z}$ as step size

$$\begin{aligned}\mathbf{x}_{giv_A} &= \mathbf{x}_{old_A} + \mathbf{F}_{r_A} \cdot \Delta\mathbf{z}\,, \\ \mathbf{x}_{giv_B} &= \mathbf{x}_{old_B} + \mathbf{F}_{r_B} \cdot \Delta\mathbf{z}\,.\end{aligned} \tag{6.13}$$

The resulting sample point $\mathbf{x}_{giv}$ is generally a non-consistent sample point. A related consistent sample point $\mathbf{x}_{cons}$ can be calculated using the algorithm described in Section 3.2.2. A consistent sample point enables the computation of the linear mapping matrices $\mathbf{F}_{r_A}$ and $\mathbf{F}_{r_B}$ for the actual sample point. These mapping matrices are used for the calculation of the next sample points, absolute and relative errors of the vector fields, and the scalar fields in the virtual state space.

Additionally, the state variables $\mathbf{x}_{cons_A}$ of circuit A are adjusted to meet the state variables of circuit B exactly. This adjustment is necessary to avoid summation of small deviations to larger errors while iteratively stepping through the state space. A detailed description of this adjustment can be found in [20, 18]. The algorithm's last step proves the equivalence of the two systems. The two systems are equal, if the calculated absolute and relative errors are smaller than an error bound for every sample point in the extended virtual state space (see Section 3.3.3).

The iterative calculation of the extended state space variables is shown in Figure 6.3 for three executions of the inner loop. At symbol ①, the verification run starts at the DC operating point for the input value of $u = 0.5$. The starting points in the original and the virtual extended state space correspond to each other by definition. After computing the mapping matrices $\mathbf{F}_{r_A}$ and $\mathbf{F}_{r_B}$, the step size Δz is determined. Now, the next state variable values in the original extended state space can be computed. In the virtual extended state space the next points are determined using Δz. The resulting points are marked with ②. The second evaluation of the inner loop results in the points labeled with ③.

3.2.2 Consistent Sample Point. In Section 2.2.2 the numerical solution of the initial value problem of a DAE system (Equation (6.4)) is given with the objective of doing transient simulation. In that section, the initial value is assumed to be a consistent initial value of the DAE system [11], which is defined as follows:

Definition 6.4 *A vector* $\mathbf{x}_{cons}$ *is a consistent initial value of Equation (6.4) if there exists a solution of Equation (6.4) that fulfills* $\mathbf{x}(t_0) = \mathbf{x}_{cons}$.

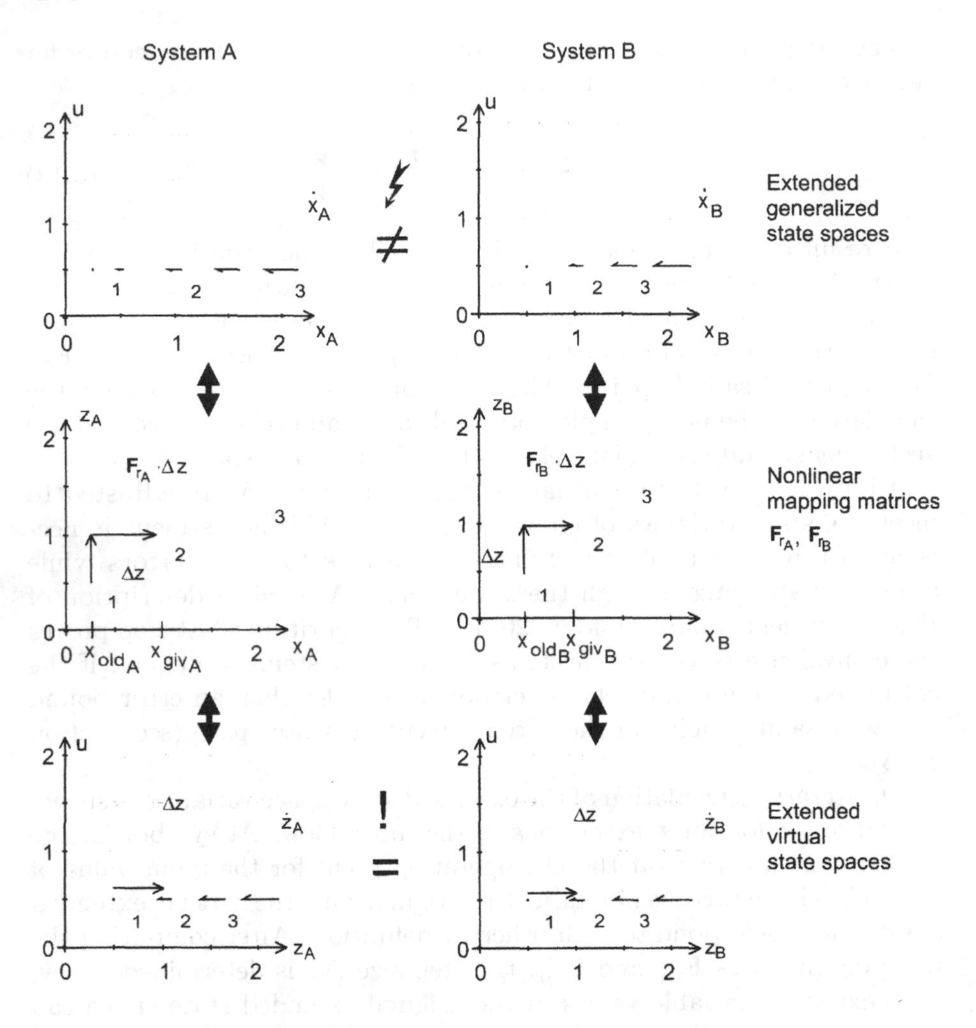

Figure 6.3. Construction of the nonlinear mapping for the example of Equation (6.11).

For this chapter, the calculation of a consistent sample point is equivalent to calculating the consistent initial value. The proposed equivalence checking approach needs consistent sample points. In this case, the estimated value calculated from Equation (6.13) can not be assumed to be a consistent sample point. That means, some state variables have slight deviations from the consistent sample point. In case of linear dependent state variables, this is caused by the nonlinearities of the DAE-system.

Therefore, the possibly inconsistent sampling point $\mathbf{x}^{\mathsf{z}}_{giv}$ has to be transfered into a consistent one $\mathbf{x}^{\mathsf{z}}_{cons}$. Additionally, we are also interested in the values of the state variables' derivatives $\dot{\mathbf{x}}^{\mathsf{z}}_{cons}$ corresponding to the consistent sample point. They will be used in error calculation procedures (see Section 3.3.3).

In order to compute the consistent sample point, the values for the state variables $\mathbf{x}^{\mathsf{z}}_{giv}$ are given. The unknown values for this problem are the variables' values $\mathbf{x}^{\mathsf{a}}_{cons}$ and the state variable's time derivatives values $\dot{\mathbf{x}}^{\mathsf{z}}_{cons}$. Hence, the following system of equations has to be solved

$$\begin{aligned} \mathbf{f}(\dot{\mathbf{x}}^{\mathsf{z}}_{cons}, \mathbf{x}^{\mathsf{z}}_{giv}, \mathbf{x}^{\mathsf{a}}_{cons}, u_{giv}) &= \mathbf{0}\,, \\ g(y_{cons}, \mathbf{x}^{\mathsf{z}}_{giv}, \mathbf{x}^{\mathsf{a}}_{cons}) &= 0\,. \end{aligned} \tag{6.14}$$

The variable's time dependencies are eliminated due to the initial value definition. Accordingly, Equation (6.14) is a system of nonlinear equations, because the state variables' derivatives $\dot{\mathbf{x}}^{\mathsf{z}}_{cons}$ are independent unknown variables. It can be solved using a Newton-Raphson solver, which should be specialized in solving equation systems of electrical circuits. The result is a consistent sample point $\mathbf{x}_{cons}$ and the state variables' derivative values $\dot{\mathbf{x}}^{\mathsf{z}}_{cons}$, representing a consistent sample point in the neighborhood of the given input vector $\mathbf{x}^{\mathsf{z}}_{giv}$.

3.3 Linear Transformation Theory

In general, calculating the nonlinear mapping functions $\mathbf{z}_A = \mathbf{t}_A(\mathbf{x}_A)$ and $\mathbf{z}_B = \mathbf{t}_B(\mathbf{x}_B)$ from the nonlinear system functions $\mathbf{f}_A$ and $\mathbf{f}_B$ directly is impossible. However, a direct mapping at a particular sample point can be calculated based on linearized systems. The three main steps in this task are

- linearizing the system in the operating point,
- computing the Kronecker canonical form, and
- calculating the mapping matrix $\mathbf{F}_r$

for both systems. The last two steps are described in detail in the following sections.

3.3.1 System Transformation to a Kronecker Canonical Form.

Starting point for the consideration is the generalized eigenvalue problem

$$(\mathbf{C} \cdot \alpha + \mathbf{G} \cdot \beta) \cdot \mathbf{x} = 0 \tag{6.15}$$

with the eigenvalues $\lambda = \frac{\alpha}{\beta}$. The matrices **G** and **C** correspond to the conductance and the capacitance matrices of the linearized system. There exist many methods to solve the generalized eigenvalue problem, e.g. the QZ algorithm.

For further studies the following assumptions are made:

- **G** is a regular and invertible matrix. Hence all eigenvalues λ are $\neq 0$. This is always true for analog systems with an existing DC operating point.
- The number of eigenvectors must be equal to the number of eigenvalues. This leads to equation systems with an index ≤ 1.

To calculate the Kronecker canonical form, two invertible transformation matrices **E** and **F** are needed, solving Equation (6.16) and Equation (6.17).

$$\mathbf{E} \cdot \mathbf{G} \cdot \mathbf{F} = \tilde{\mathbf{G}} = \begin{bmatrix} -\lambda_1 & 0 & \cdots & 0 \\ 0 & -\lambda_2 & & \vdots \\ \vdots & & \ddots & 0 \\ 0 & \cdots & 0 & 1 \end{bmatrix} \tag{6.16}$$

$$\mathbf{E} \cdot \mathbf{C} \cdot \mathbf{F} = \tilde{\mathbf{C}} = \begin{bmatrix} 1 & 0 & \cdots & 0 \\ 0 & 1 & & \vdots \\ \vdots & & \ddots & 0 \\ 0 & \cdots & 0 & 0 \end{bmatrix} \tag{6.17}$$

For systems with an index ≤ 1 the matrices $\tilde{\mathbf{G}}$ and $\tilde{\mathbf{C}}$ are diagonal matrices. The upper part of the diagonal of matrix $\tilde{\mathbf{G}}$ consists of the negative eigenvalues $-\lambda_1 \cdots - \lambda z$, the lower part of the diagonal is filled with ones. A similar structure applies to matrix $\tilde{\mathbf{C}}$. The upper part of the diagonal is filled with ones and the lower part of the diagonal with zeros. In both matrices the upper part represents the finite eigenvalues, the lower part stands for the infinite eigenvalues. With these two matrices we get the Kronecker canonical form

$$s \cdot \tilde{\mathbf{C}} \cdot \mathbf{x} + \tilde{\mathbf{G}} \cdot \mathbf{x} = 0\,. \tag{6.18}$$

The complex frequency variable s results from the Laplace transformation of the linearized system shown in Equation (6.8). The results from the generalized eigenvalue problem are the eigenvalues and the right and left eigenvectors. If we arrange the right eigenvectors columnwise in a matrix $\mathbf{V}_r$, the solution of the generalized eigenvalue problem can be written in matrix form:

$$\mathbf{C} \cdot \mathbf{V}_r \cdot -\tilde{\mathbf{G}} + \mathbf{G} \cdot \mathbf{V}_r \cdot \tilde{\mathbf{C}} = 0\,. \tag{6.19}$$

The transformation matrices $\mathbf{E}$ and $\mathbf{F}$ can be determined by coefficient comparison. Therefore, Equation (6.17) has to be multiplied with $\mathbf{E}^{-1}$ from the left side and Equation (6.19) has to be multiplied with $\tilde{\mathbf{G}}^{-1}$ from the right side. Because both $\tilde{\mathbf{G}}^{-1}$ and $\tilde{\mathbf{C}}$ are diagonal matrices, they can be swapped. Finally, the results of the coefficient comparison are

$$\begin{aligned} \mathbf{F} &= \mathbf{V}_r\,, \\ \mathbf{E} &= \tilde{\mathbf{G}} \cdot \mathbf{V}_r^{-1} \cdot \mathbf{G}^{-1}\,. \end{aligned} \tag{6.20}$$

$\mathbf{E}$ and $\mathbf{F}$ are defined, so that the Kronecker canonical form, shown in März [27] can be calculated. Another method for the determination of the transformation matrices can be found in [10] (elimination method). Note, that for systems with an index > 1 these methods can not be used. Instead, a different approach has to be chosen.

3.3.2 DAE System Transformation into the Virtual State Space. In order to compute a canonical form, the linearized systems shown in Equation (6.21) have to be transformed into an explicit state space description [30] Equation (6.22)

$$\begin{aligned} s \cdot \mathbf{C} \cdot \mathbf{x} + \mathbf{G} \cdot \mathbf{x} &= \mathbf{q} \cdot u\,, \\ y &= \mathbf{r}^T \cdot \mathbf{x}\,, \end{aligned} \tag{6.21}$$

$$\begin{aligned} \dot{\mathbf{z}} &= \mathbf{A}_D \cdot \mathbf{z} + \mathbf{b}_D \cdot u\,, \\ y &= \mathbf{c}_D^T \cdot \mathbf{z} + d_D \cdot u\,. \end{aligned} \tag{6.22}$$

The transformation from the virtual state space into the state space is defined as

$$\mathbf{x} = \mathbf{F}_r \cdot \mathbf{z}\,, \tag{6.23}$$

where $\mathbf{z} \in \mathbb{R}^z$ is the state vector in the virtual state space and $\mathbf{F}_r \in \mathbb{R}^{n \times z}$ is the finite part from the transformation matrix $\mathbf{F}$, shown in Equation (6.20).

To derive the canonical form the state space is expanded by the infinite state space variables $\mathbf{z}_\infty$. This theoretical consideration is needed to calculate the unknown matrix $\mathbf{A}_D$, the vectors $\mathbf{b}_D$ and $\mathbf{c}_D^T$ as well as the scalar d. Using these extensions, Equation (6.23) can be written as

$$\mathbf{x} = \left[\, \mathbf{F}_r \;\middle|\; \mathbf{F}_\infty \,\right] \cdot \left[\frac{\mathbf{z}}{\mathbf{z}_\infty}\right] = \mathbf{F} \cdot \mathbf{z}^*\,. \tag{6.24}$$

If Equation (6.21) is multiplied with $\mathbf{E}$ from the left side and furthermore, if the variables $\mathbf{x}$ are replaced by the transformation (6.24), the following expressions can be derived:

$$\begin{aligned} s \cdot \mathbf{E} \cdot \mathbf{C} \cdot \mathbf{F} \cdot \mathbf{z}^* + \mathbf{E} \cdot \mathbf{G} \cdot \mathbf{z}^* &= \mathbf{E} \cdot \mathbf{q} \cdot u\,, \\ y &= \mathbf{r}^T \cdot \mathbf{F} \cdot \mathbf{z}^*\,. \end{aligned} \tag{6.25}$$

With the definition of the matrices $\tilde{\mathbf{G}}$ and $\tilde{\mathbf{C}}$ and dividing Equation (6.25) into an upper and a lower part, we get

$$s \cdot \begin{bmatrix} \mathbf{I} & \mathbf{0} \\ \mathbf{0} & \mathbf{0} \end{bmatrix} \begin{bmatrix} \mathbf{z} \\ \hline \mathbf{z}_\infty \end{bmatrix} + \begin{bmatrix} -\mathbf{\Lambda} & \mathbf{0} \\ \mathbf{0} & \mathbf{I} \end{bmatrix} \begin{bmatrix} \mathbf{z} \\ \hline \mathbf{z}_\infty \end{bmatrix} = \begin{bmatrix} \mathbf{E}_r \\ \hline \mathbf{E}_\infty \end{bmatrix} \cdot \mathbf{q} \cdot u\,, \tag{6.26}$$

$$y = \mathbf{r}^T \cdot \left[\, \mathbf{F}_r \;\middle|\; \mathbf{F}_\infty \,\right] \cdot \begin{bmatrix} \mathbf{z} \\ \hline \mathbf{z}_\infty \end{bmatrix}\,. \tag{6.27}$$

Thereby $\mathbf{\Lambda}$ represents the finite eigenvalues, arranged in a diagonal matrix. The results can be determined using the upper part of Equation (6.26) and a rewritten form of Equation (6.27), combined with the lower part of Equation (6.26)

$$\begin{aligned} \mathbf{A}_D &= \mathbf{\Lambda}\,, \\ \mathbf{b}_D &= \mathbf{E}_r \cdot \mathbf{q}\,, \\ \mathbf{c}_D^T &= \mathbf{r}^T \cdot \mathbf{F}_r\,, \\ d_D &= \mathbf{r}^T \cdot \mathbf{F}_\infty \cdot \mathbf{E}_\infty \cdot \mathbf{q}\,. \end{aligned} \tag{6.28}$$

Before the calculation of the virtual state space variables is performed, the eigenvalues are sorted according to their magnitude. This is necessary to obtain a unique mapping of each state space variable from one system to the other.

The resulting systems can have different orders. The number of eigenvalues of system A z_A and the number of eigenvalues of system B z_B often differs. In this case, the number of eigenvalues has to be reduced to the minimum number of eigenvalues of both systems

$$z_r = min(z_a, z_b)\,. \tag{6.29}$$

Additionally, the number of used eigenvalues z_r can be limited by the user. This technique is comparable to the dominant poles model order reduction. An alternative solution would be to reduce both systems using other model order reduction techniques [13, 12].

Different scales of the eigenvectors are eliminated by using the vectors $\mathbf{b}_D$ and the vectors $\mathbf{c}_D^T$ for an additional scaling condition. The scaling problem results from the fact, that the eigenvectors $\mathbf{V}_r$, which are used to calculate the transformation matrices **E** and **F**, have only fixed directions but not fixed lengths. Therefore, an additional scaling matrix $\mathbf{T}_z$ is needed

$$\begin{aligned} \mathbf{A}_{D_A} \cdot \mathbf{T}_z &\stackrel{!}{=} \mathbf{T}_z \cdot \mathbf{A}_{D_A} , \\ \mathbf{b}_{D_A} &\stackrel{!}{=} \mathbf{T}_z \cdot \mathbf{b}_{D_B} , \\ \mathbf{T}_z \cdot \mathbf{c}_{D_A}^T &\stackrel{!}{=} \mathbf{c}_{D_B}^T . \end{aligned} \tag{6.30}$$

To calculate the coefficients of the scaling matrix $\mathbf{T}_z$ the three Equations (6.30) are combined to one equation system. This overdetermined equation system can be solved using a least square approach. Afterwards, the scaling matrix $\mathbf{T}_z$ is used to scale the mapping matrices

$$\begin{aligned} \mathbf{F}'_{r_A} &= \mathbf{T}_z \cdot \mathbf{F}_{r_A} , \\ \mathbf{F}'_{r_B} &= \mathbf{F}_{r_B} . \end{aligned} \tag{6.31}$$

As shown in Equation (6.31), only the mapping matrix $\mathbf{F}_A$ is affected by the scaling algorithm. The mapping matrix $\mathbf{F}_B$ remains unchanged. The complete algorithm to calculate the mapping matrices is shown in Figure 6.4.

Determine linear mapping matrices $\mathbf{F}_{r_A}$ and $\mathbf{F}_{r_B}$() {
  linearize systems A and B at the sample point.
  calculate the eigenvalues and the eigenvectors
  sort eigenvalues and eigenvectors according to their magnitude
  if $z_a \neq z_b$ **then** {
    reduce the system A and B to the order of z_r
  }
  calculate transformation matrices $\mathbf{E}_A$, $\mathbf{F}_A$, $\mathbf{E}_B$ and $\mathbf{F}_B$.
  calculate the state space descriptions:

$$\begin{aligned} \dot{\mathbf{z}}_\mathbf{A} &= \mathbf{A}_{D_A} \cdot \mathbf{z}_A + \mathbf{b}_{D_A} \cdot u_A \\ y_A &= \mathbf{c}_{D_A}^T \cdot \mathbf{z}_A + d_{D_A} \cdot u_A \\ \dot{\mathbf{z}}_\mathbf{B} &= \mathbf{A}_{D_B} \cdot \mathbf{z}_B + \mathbf{b}_{D_B} \cdot u_B \\ y_B &= \mathbf{c}_{D_B}^T \cdot \mathbf{z}_B + d_{D_B} \cdot u_B \end{aligned}$$

  scale mapping matrices with $\mathbf{T}_z$
  return $\mathbf{F}'_{r_A}$ and $\mathbf{F}'_{r_B}$
}

Figure 6.4. Algorithm for linear transformation generation.

3.3.3 Error Calculation. The derivatives $\dot{\mathbf{z}}_A$, $\dot{\mathbf{z}}_B$ have to be calculated in order to determine the errors between the state variables in the virtual state space. The computation of the consistent sample point $\mathbf{x}_{cons}$ also calculates the time derivatives $\dot{\mathbf{x}}^{\mathsf{z}}_{cons}$ (see Section 3.2.2). Using Equation (6.23) we could transform the $\dot{\mathbf{x}}$ vector into the virtual state space. Since we do not have the complete $\dot{\mathbf{x}}$ vector, a reduced equation system has to be build. Multiplying Equation (6.23) from the left side with $\mathbf{C}$ and building the time derivative leads to the following adequate reduced system

$$\mathbf{C} \cdot \dot{\mathbf{x}}_{cons}|_{\dot{\mathbf{x}}^{\mathsf{a}}_{cons}=0} = \mathbf{C} \cdot \mathbf{F} \cdot \dot{\mathbf{z}}\,. \tag{6.32}$$

The part of the vector $\dot{\mathbf{x}}_{cons}$ which belongs to the algebraic variables $\dot{\mathbf{x}}^{\mathsf{a}}_{cons}$ is set to zero. Due to the structure of $\mathbf{C}$, the solution of Equation (6.32) will not depend on $\dot{\mathbf{x}}^{\mathsf{a}}_{cons}$. Equation (6.32) can be solved with a least square algorithm for $\dot{\mathbf{z}}$ where the residuum would be in general close to zero. Computing derivatives of virtual state variables for both systems enables the error calculation.

We receive the errors $\varepsilon_{\dot{\mathbf{z}}}$ and ε_y straightforward by calculating the differences $\varepsilon_{\dot{\mathbf{z}}} = \|\dot{\mathbf{z}}_A - \dot{\mathbf{z}}_B\|$ and $\varepsilon_y = \|y_A - y_B\|$ for each sample point. Additionally, relative and mean relative errors are calculated.

3.4 Experimental Results

In this section, the described equivalence checking approach is applied to two nonlinear examples. The first example, a Schmitt trigger circuit, is highly nonlinear. The model checking approach described in Section 4 also deals with this example (Section 4.3.1). The second example, a bandpass circuit consists of more state variables and is modeled on transistor level.

3.4.1 Schmitt Trigger Example. The netlist of the inverting Schmitt trigger consists of an opamp (operational amplifier) behavioral model, two resistors, and an output capacitance as shown in Figure 6.5. The opamp's open loop gain is 10000. The output voltage restriction is ± 5 V and the maximum output current is 80 mA. The capacitor is set to 1 μF, resistors R_1 and R_2 are both set to 10 kΩ. Thus, the switching threshold is about ± 2.5 V and the output voltage varies between $+5$ V and -5 V.

This circuit has only one state variable, namely, the output voltage V_{out}. Additionally, it has one input signal V_{in}. To consider all states that might occur in the circuit, the extended state space is chosen to be $V_{out} = [-7\ V\ ..\ 7\ V]$ and $V_{in} = [-7\ V\ ..\ 7\ V]$.

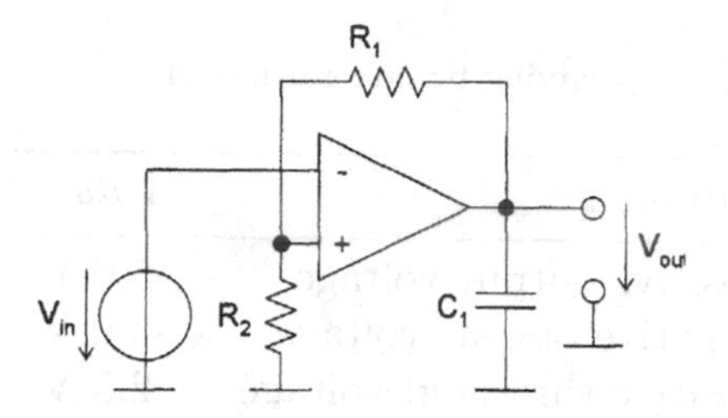

Figure 6.5. Schmitt trigger circuit.

A vector plot of the state variable derivatives $\dot{V}_{out}$ is shown in Figure 6.6. The stable and unstable equilibriums are marked with solid and dashed lines, respectively. There are two stable equilibriums states at 5.0 V and −5.0 V with in the input voltage range of $V_{in} = -2.5$ V to 2.5 V.

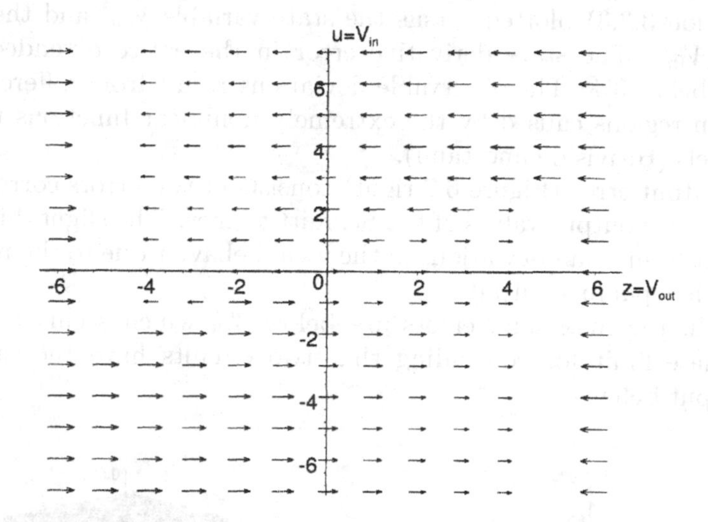

Figure 6.6. State variables derivatives $\dot{V}_{out}$ for Schmitt trigger circuit.

The specification for this example is a behavioral model of the Schmitt trigger function. It is written in an analog hardware description language. The parameters for the model are shown in Table 6.1. The model consists of one state variable and nonlinear equations based on Heaviside functions modeling the Schmitt trigger's behavior. The time constant for the state variable is determined by estimating the time constant resulting from the capacitance C_1 and the output current limitation of the opamp.

Table 6.1. Parameters of specifying behavioral model.

Parameter	*Value*
Positive output voltage	5.0 V
Negative output voltage	-5.0 V
Positive threshold voltage	2.5 V
Negative threshold voltage	-2.5 V
Switching time constant	6.7 ms

A first equivalence checking run verifies the behavioral model versus the circuit. The step size was chosen to 0.2 V in the extended state space. 9179 points are calculated in about 189 s on a SUN SPARC Ultra 10. The results are shown in Figure 6.7. The errors are mean relative errors (see Section 3.3.3) plotted versus the state variable V_{out} and the input variable V_{in}. The state derivative error in the entire extended state space is below 3%. The observable deviations result from differences in transition regions caused by the extremely nonlinear functions used in the models (Heaviside and tanh).

The output error (Figure 6.7 right) consists of two errors corresponding to the two output values of the Schmitt trigger. The slight difference is a result from small deviations in the ideal behavior due to the resistive load at the opamp's output.

Overall, the maximum errors are below 3% which seems to be an appropriate limit for concluding that two circuits have the same input/output behavior.

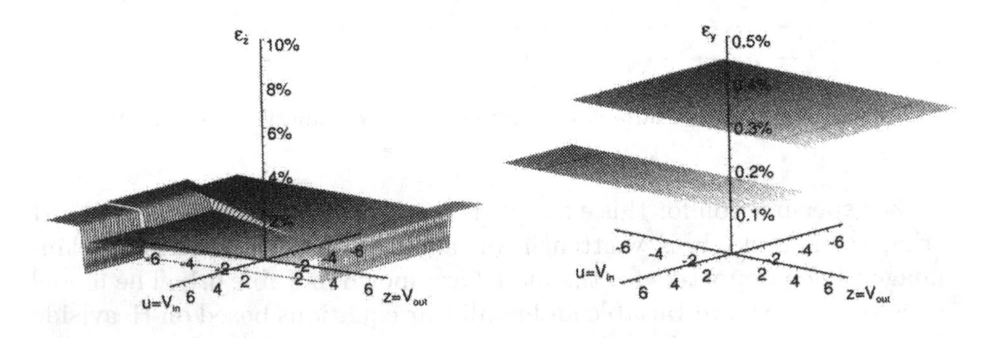

Figure 6.7. Error of derivatives of state variables $\varepsilon_{\dot{z}}$ and output error ε_y for Schmitt trigger circuit.

In order to show the equivalence checking results in case of a faulty circuit behavior, we change the resistor R_1 to 8 kΩ. The new system primarily differs in the state variable derivatives. The plot of its mean relative error is shown in Figure 6.8. The modified resistor leads to a displacement of the region corresponding to the positive and negative output voltage. In Figure 6.8 two of those regions can be identified. The first one is the triangular area indicating different negative threshold voltages. The second one is a difference in the positive threshold voltage resulting in a rectangular plane of large errors parallel to the V_{out}, V_{in} plane. The maximum error is 86% clearly indicating a different behavior.

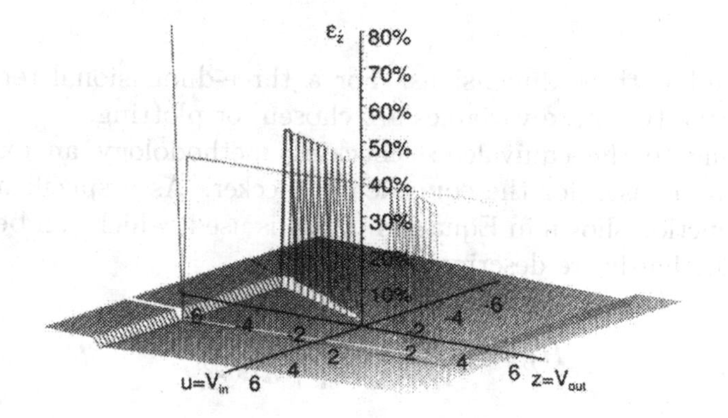

Figure 6.8. State variable derivative error $\varepsilon_{\dot{z}}$ for the faulty Schmitt trigger circuit.

3.4.2 Bandpass Example. The second example is a Sallen-Key bandpass filter. This circuit consists of five resistors, two capacitors, and an opamp (see Figure 6.9). The opamp itself is modeled with eight MOS transistors using a full BSIM3 model. The resistors R_1 and R_2 are set to 15.9 $k\Omega$, the resistor R_3 is set to 31.8 $k\Omega$, R_4 is set to 5 $k\Omega$, and finally R_5 is set to 10 $k\Omega$. The capacitors C_1 and C_2 are both set to 10 nF

The circuit has two dominant state variables: the nodal-voltages V_I and V_{II} and seven parasitic state variables resulting from the capacitances inside the MOS transistors. Since the two dominant state variables correspond to a linear combination of the two nodal voltages, a clear mapping of the virtual state variables **z** to the nodal voltages is impossible. Additionally, the number of states have to be reduced in order to compare the circuit with the given specification, which covers the dominant states only. Together with the input signal V_{in} the extended

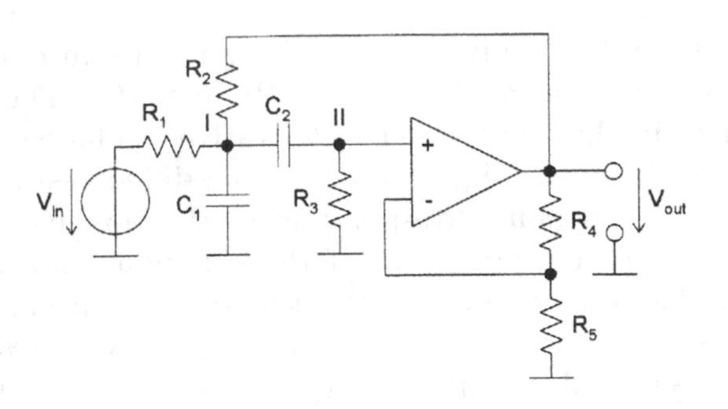

Figure 6.9. Bandpass circuit.

state space has three dimensions. For a three-dimensional representation, only the two state variables are chosen for plotting.

According to the equivalence checking methodology an executable specification is used for the equivalence checker. As a specification the transfer function shown in Equation (6.33) is used, which can be written in an analog hardware description language

$$H_{BP} = \frac{b_1 \cdot s_n}{1 + a_1 \cdot s_n + a_2 \cdot s_n^2} \,. \tag{6.33}$$

The coefficients are chosen as $a_1 = 0.35775\ ms$, $a_2 = 0.025281\ \mu s^2$ and $b_1 = 0.11925\ ms$. The transfer function has two conjugated complex poles.

The verification run compares the bandpass netlist versus the transfer function model. In the first run the boundaries of the extended state space are set to $[-0.9\ V\ ..\ 0.9\ V]$. The input value is set to a constant value of $1.0\ V$, the step size is $0.02\ V$. The algorithm needs 236 seconds to compare 8281 points in the virtual state space. The results are shown in Figure 6.10. The mean relative errors are plotted over the two virtual state space variables z_0 and z_1.

In contrast to expectations, the maximum errors are quite large. For the derivatives of the state variables a maximum error of over 100% is calculated. The output error is 6%, which exceeds the postulated error bound. The derivatives of the state variables are similar for both systems, as long as the state variables do not exceed its limits. These limits result from the nonlinearities inside the transistor models not considered by the linear transfer function.

To illustrate this effect, a second run was invoked with reduced boundaries. Now the mapping of the state space variables is done in the range of $[-0.45\ V\ ..\ 0.45\ V]$ with a step size of $0.01\ V$. This leads to 8281

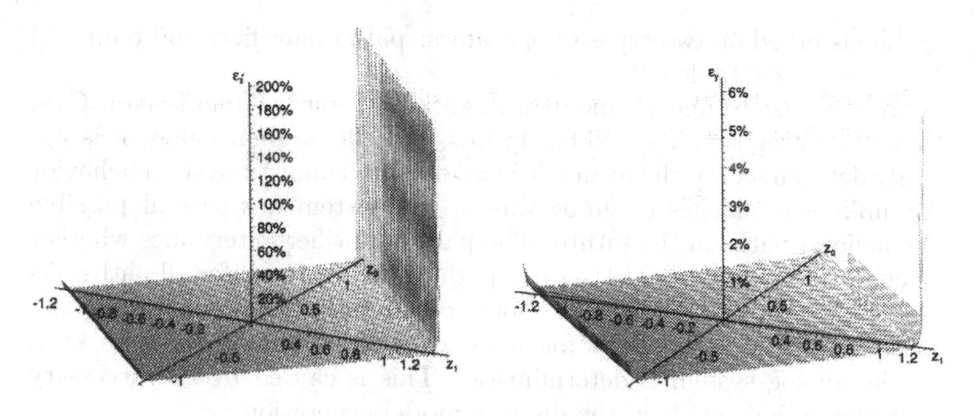

Figure 6.10. State variables derivative error $\varepsilon_{\dot{z}}$ and output error ε_y for the Bandpass circuit.

calculated points in the virtual state space. The maximum error for the derivatives of the state variables is now only 0.05% and the maximum output error is 0.08%. Inside these state space boundaries we can assume, that the equivalence of both systems regarding the input/output behavior is proven.

4. Model Checking

Model checking algorithms prove or to disprove the correctness of a given specification property of an actual system. As we have shown in Section 2.2, the state space description represents the entire behavior of a nonlinear dynamic system. To enable a model checking procedure for nonlinear dynamic systems, the specification properties have to be described as properties of the state space description. These state space properties will be checked versus the system's state space [17, 16].

The next section introduces a property description language enabling the definition of analog system properties within the state space description. The main algorithmic task is to check these state space properties versus the system description. This issue will be addressed in the following four sections. Finally, some experimental results are given, showing the use of the developed prototype.

4.1 Model Checking Language

One of the most common language in digital model checkers is CTL (Computation Tree Logic) invented by Clarke and Emerson [6]. The language is based on two types of operators: path quantifiers and temporal operators (see Table 6.2).

In contrast to linear time logics like LTL (Linear Time Logic), CTL is a branching time logic. That means, that the system behavior is not fully determined for the future. Due to nondeterministic system behavior or unknown changes of input values, the system has several possible behavioral paths in the future. The path quantifier determines whether a condition is valid for at least one path in the future or for all paths. As we will see in the following sections, consideration of nondeterministic system behavior is essential for model checking on analog systems, even if the analog system is deterministic. This is caused by the necessary approximations made in the discrete model generation.

The temporal operator is used to describe the time dependent system behavior. Since this approach is focused on dynamical systems, the time dependencies have to be included in the analog property description.

Obviously, CTL in its basic form is not suited for the description of analog properties because it considers only boolean state variables. Thus, the language has been extended by a minimum set of operations enabling its use for analog models. Table 6.2 gives a short syntax overview on the CTL language and the analog extensions (bold symbols in the table). The extended language is called CTL-A.

For example, the equation $\Theta = \mathsf{AF}(state_1)$ can be read as follows: All paths starting in a state within Θ will eventually reach a state in which $state_1$ is true. To simplify matters, we do not distinguish between a CTL condition and the set of states that fulfills this condition, both will be named by capital Greek letters.

Since the domain of digital state variables is restricted to boolean values, the statements a and $\neg a$ cover the whole domain. Analog state variables are defined continuously. Therefore, the greater and smaller operators are introduced in the CTL-A language definition. In this way, half planes can be described in the continuous state space e.g. $x_1 > -13.2546$. The combination of several half planes with boolean operators enables the definition of arbitrary Manhattan polytopes in a continuous n-dimensional space. Boolean variables are left out in the following model checking algorithms because only purely analog systems are considered. However, an extension of this approach to mixed signal or hybrid systems is possible.

The following example explains the use of CTL-A to describe transient system properties. More complex CTL-A equations can be found in Section 4.3. The system behavior is given by $\{\dot{\mathbf{x}} = (1,0)^T\}$. Figure 6.11 shows the system dynamics as light gray arrows in the state space. The CTL-A equation to be analyzed is given in Equation (6.34)

Table 6.2. CTL-A syntax description.

$\Phi := a \mid \mathbf{z} * \mathbf{v} \mid \Phi \circ \Phi \mid \neg\Phi \mid \rhd \diamond \Phi \mid \rhd \Phi \, \mathsf{U} \, \Phi \mid \rhd \diamond^{-1} \Phi \mid \rhd \Phi \, \mathsf{U}^{-1} \Phi$		
a	boolean state variables	
$\mathbf{z}$	**continuous state variables**	
$\mathbf{v}$	**constant real values**	
$*$	**analog operators**	$>$ = **greater**
		$<$ = **smaller**
$\circ$	boolean operators	$\vee$ = or
		$\wedge$ = and
$\neg$		$\neg$ = not
$\rhd$	path quantifiers	A = on all paths
		E = on some paths
$\diamond$	temporal operators	X = next-time
		F = eventually
		G = always
U		U = until
$^{-1}$	past	time inversion

$$\begin{aligned} \Phi &= (x_{(1)} > 2) \;\wedge\; (x_{(1)} < 3) \;\wedge\; (x_{(2)} > 1) \;\wedge\; (x_{(2)} < 3) \\ \Psi &= \mathsf{EF}\,(\Phi)\,. \end{aligned} \tag{6.34}$$

Since all states at the left side of Φ have a path that will eventually reach Φ, it is obvious that Ψ is true for these states. The result is shown as the gray region in Figure 6.11.

In the classical CTL definition, the time model is restricted to the future. It has been shown in some experimental results [17], that consideration of the past is essential for analog specification properties. The time model for the past is branching and infinite. Thus, the operator $^{-1}$ inverses all system transitions with respect to time. For example $\mathsf{EF}^{-1}(x)$ means, that the condition x must have been valid on at least one path in the past. A collection of other past time definitions and languages can be found in [26]. For the system described above the formula $\mathsf{EF}^{-1}\,(\Phi)$ results in the region on the right hand side of Φ.

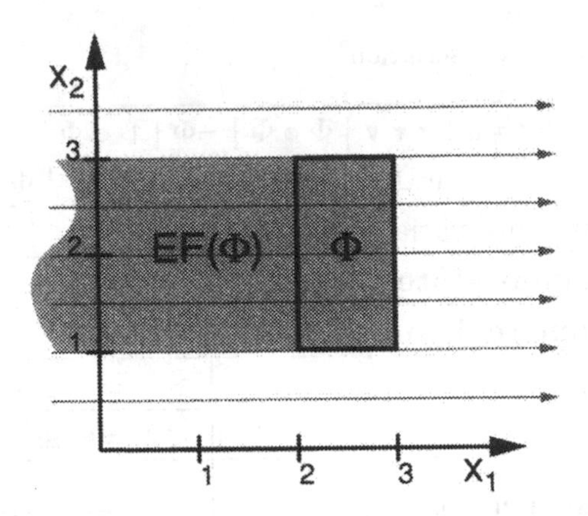

Figure 6.11. Result of the CTL-A formula: EF (Φ).

The analog specifications that can be described in CTL-A are limited. For example, frequency domain behavior is not covered by the language. For future development a further extension of the property description language is necessary to focus on the needs in analog design. However, it will be shown below that checking a variety of analog properties is possible even with this minimum extension of digital CTL.

4.2 Analog Model Checking Algorithm

We have shown in the last section that all analog system properties covered by CTL-A can be represented as regions in the continuous state space. The analog model checking task is to calculate the region fulfilling a given CTL-A formula.

For a numerical analysis, the continuous variables in an analog system - state values and time - have to be transfered into a discrete state space description with state transition relations. The next sections illustrate this process.

4.2.1 Transition Systems. Digital and hybrid model checking tools are often based on transition systems.

Definition 6.5 *A state transition system $T = (\mathcal{Q}, \mathcal{Q}_0, \sum, R)$ consists of*

- *a set of states $\mathcal{Q}$,*
- *a set of initial states $\mathcal{Q}_0$,*
- *a set of generators or events $\sum$, and*
- *a state transition relation $R \subseteq \mathcal{Q} \times \sum \times \mathcal{Q}$.*

An analog system description (Equation (6.4)) is also a state transition system. The set of states $\mathcal{Q}$ can be represented by the continuous extended state space $\mathbb{R}^n$. The number of states $x \in \mathcal{Q}$ is infinite, due to the continuous definition of the state variables. The initial state $\mathcal{Q}_0$ is a single point or a region in the state space. Often, but not necessarily, this is the DC operating point. There exist only $i+1$ generators $\sum$ causing state transitions, namely, the time t and the i input values $\mathbf{u}(t)$. The state transition relation $R \subseteq \left(\mathbb{R}^n \times \mathbb{R}^{i+1} \times \mathbb{R}^n\right)$ is a continuous function given by the time derivatives $\dot{\mathbf{x}}^{\mathbf{z}}(t)$. We assume the state variables $\mathbf{x}^{\mathbf{z}}$ to be independent. For systems with dependent state variables (see Section 2.2) the algorithms have to be modified by taking into account approaches from [4] and Section 3.2.2. Since the output function $g\left(y(t), \mathbf{x}(t)\right)$ in Equation (6.4) is very simple in most practical cases, it will be assumed that output values are only state variables $\mathbf{x}$. Thereby, function g can be neglected for the rest of this chapter.

The actual state transition can be calculated by integrating the differential equation system (shown for the ODE case where $\mathbf{x} = \mathbf{x}^{\mathbf{z}}$)

$$\mathbf{f}_{int}(\mathbf{x}(t_0), \mathbf{u}(t), \delta t) = \\ \mathbf{x}(t_0) + \int_{t_0}^{t_0+\delta t} \left\{\dot{\mathbf{x}}(\tau) \mid \mathbf{f}\left(\dot{\mathbf{x}}(\tau), \mathbf{x}(\tau), \mathbf{u}(\tau)\right) = 0\right\} \, d\tau. \tag{6.35}$$

Equation (6.35) has no direct time dependency but it depends on the input signals $\mathbf{u}(t)$. Thus, the generators $\sum$ are not time t and input values but rather a time difference δt and the input values. Without losing generality, the time difference δt might be either infinitesimal small or a finite value.

Thus, digital and analog systems can be described by transition systems. However, the analog system is continuous in time and state variable values. Therefore, a method has to be developed to approximate this behavior numerically. Some of the following algorithms have been inspired by research in the area of approximating dynamical behavior [8]. Despite the similarities, there are differences, mainly caused by the overall target of the algorithms. The work of Kurshan and McMillan [25] is also linked to the following algorithms.

4.2.2 Discrete Time Steps. In Section 4.2.1 we found that the transition relation $R \subseteq \left(\mathbb{R}^n \times \mathbb{R}^{i+1} \times \mathbb{R}^n\right)$ for an analog system is given as a continuous function and the actual state transition can be calculated by integrating this function. According to Section 2.2.2 Equation (6.35) can be solved using numerical integration with a small time step $\Delta t = t_n - t_{n-1}$

$$\mathbf{f}_{num}\left(\mathbf{x}(t_{n-1}), \mathbf{u}(t_n), \Delta t\right) = \left\{ \mathbf{x}(t_n) \,\middle|\, \mathbf{f}\left(\frac{\mathbf{x}(t_n)-\mathbf{x}(t_{n-1})}{\Delta t}, \mathbf{x}(t_n), \mathbf{u}(t_n)\right) = \mathbf{0} \right\}. \tag{6.36}$$

Besides numerical problems, an error due to the finite length of Δt can not be avoided. To bound this error, a local step size control mechanism is necessary. The algorithm used is known from transient analog simulations. It takes the second derivative with respect to time for a local measurement of the integration error. If the given error threshold is exceeded, the step size is reduced. Otherwise, the transient step is accepted. This method can be used directly in the analog model checking tool. An arbitrary test point $\mathbf{x}^{\mathbf{z}}(t)$ in the state space is mapped to its successor state $\mathbf{x}^{\mathbf{z}}(t + \Delta t)$, depending on the actual step size Δt and the input signals $\mathbf{u}(t)$. In contrast to transient simulation, there is no temporal predecessor state for a test point. A second time step has to be calculated for each point to determine the second derivative, enabling a local step size control.

In general, the time step Δt will vary throughout the state space, due to the step size control. As we will see later, this makes the checking of explicit time dependencies difficult because Δt has to be stored for each transition separately. To reduce the problem, the time step is chosen to be equal for each point within one state space region (to be defined in Section 4.2.3). Kurshan and McMillan [25] proposed a constant time step Δt for the whole space developed by several small numerical integration steps (segments of trajectories). Despite the advantage of a constant time step, this approach is not suitable for all circuits since the step size variation in terms of state variable values may be large throughout the state space.

Thus, every state space point $\mathbf{x}^{\mathbf{z}}(t)$ can be mapped to its successor point $s(\mathbf{x}^{\mathbf{z}}(t)) = f_{num}(\mathbf{x}(t), \mathbf{u}_{const}, \Delta t_{length\_control})$ (only state variables $\mathbf{x}^{\mathbf{z}}(t)$ are taken into account) including a local step length control and assuming given input values. The resulting tuple of test and target point is represented by a successor vector $sv(\mathbf{x}^{\mathbf{z}}(t)) = s(\mathbf{x}^{\mathbf{z}}(t)) - \mathbf{x}^{\mathbf{z}}(t)$ in the state space.

4.2.3 State Space Subdivision. To get a discrete and finite state model, the continuous and infinite state space has to be bounded and subdivided by rectangular boxes. In general, boxes are not necessarily the best choice [21]. However, for implementation reasons boxes are the far most convenient data structure. Other subdivision geometries might be considered during future improvements. The restriction to a finite region will also be discussed in Section 4.2.5.

Since there is no digital environment, a natural subdivision for the starting region, given for example by threshold values of digital state transitions, is missing [1]. Furthermore, to retain the analog system behavior correctly, a sufficient number of subdivisions is necessary, especially in state space regions with highly nonlinear behavior. This differs from approaches focusing on digital circuit behavior [25]. However, the number of discrete regions should be kept as small as possible to reduce the total runtime.

The algorithm starts with a user controlled uniform subdivision in all state space dimensions. Then, an automatic subdivision strategy is used to react on different system dynamics, depending on the actual state space region. The main target is to get a uniform behavior in each state space box. The uniformity is measured by the variation of the successor vectors (sv) starting at given test points P_{test} within the state space regions (see Section 4.2.2 and 4.2.4). Namely, vector length variation l_{var} and angle a_{var} between different vectors are considered. Equations (6.37) and (6.38) give the definitions of these values. The function l gives the length of the argument vector or vector component. Input value variations are not taken into account

$$l_{var} = 1 - \frac{\min_{\mathbf{x}^{\mathbf{z}} \in \mathcal{P}_{test}} l(sv(\mathbf{x}^{\mathbf{z}}))}{\max_{\mathbf{x}^{\mathbf{z}} \in \mathcal{P}_{test}} l(sv(\mathbf{x}^{\mathbf{z}}))}. \tag{6.37}$$

To simplify matters, the maximum angle between the successor vectors is not calculated in detail. The angle variation is approximated by the cosine of the vectors with respect to each state space dimension (Equation (6.38))

$$a_{var} = \max_{k \in n} \left\{ \max_{\mathbf{x}^{\mathbf{z}} \in \mathcal{P}_{test}} \frac{l(sv(\mathbf{x}^{\mathbf{z}})_{(k)})}{l(sv(\mathbf{x}^{\mathbf{z}}))} - \min_{\mathbf{x}^{\mathbf{z}} \in \mathcal{P}_{test}} \frac{l(sv(\mathbf{x}^{\mathbf{z}})_{(k)})}{l(sv(\mathbf{x}^{\mathbf{z}}))} \right\}. \tag{6.38}$$

Box subdivision is continued recursively until l_{var} and a_{var} drop under a given threshold or a given subdivision depth is exceeded.

Within the expected accuracy, all boxes fulfilling the l_{var} and a_{var} thresholds do not contain fix points. The reason is that fix points are always surrounded by regions with nonuniform behavior in terms of Equations (6.37) and (6.38). The fix point information is stored and used in the transition relation algorithm (Section 4.2.4).

Additional subdivisions are applied if the successor vectors in a region are too short in relation to the box size. These subdivisions are mainly necessary in regions where the system is strongly nonlinear, which implies Δt to be very small. Each box in the state space represents a single state in the discrete model. Thus, the set of i states is given by $\mathcal{Q} = \{box_1, box_2, ..., box_i\}$.

Until now, symbolic state representation techniques known form digital model checking have not been used because the usability of these methods is not obvious for analog problems.

4.2.4 Transition Relation. The last step in getting a discrete system model is the transition relation between state space regions. In Section 4.2.2 successor points for single state space points have been defined. Using this point to point relation (s), the target Region $\mathcal{R}_{exact}$ is given by the set of all target points associated with a test point within the state space Region $\mathcal{R}_{test}$ (see Equation (6.39)), as illustrated by the gray regions in Figure 6.12

$$\mathcal{R}_{exact} = \{s(\mathbf{x}^z) \mid \mathbf{x}^z \in \mathcal{R}_{test}\} = s(\mathcal{R}_{test}). \tag{6.39}$$

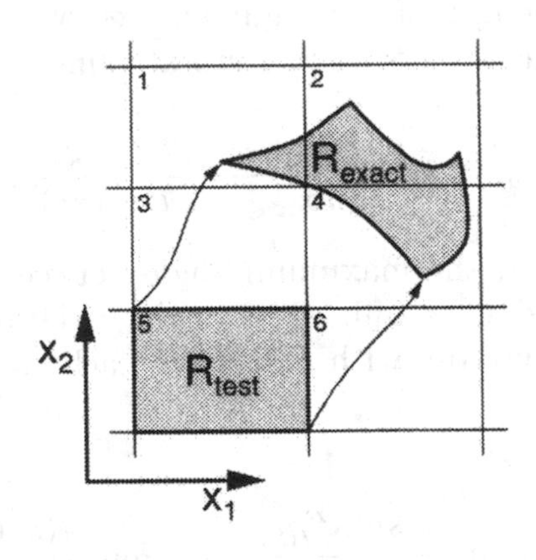

Figure 6.12. Exact nonlinear transformation of Region $\mathcal{R}_{test}$.

One way to approximate the target region is to choose a number of test points $\mathcal{P}_{test}$ within the test region (e.g. randomly, grid based, or corner values) and to calculate the dedicated target points $\mathcal{P}_{target}$. The target Region $\widetilde{\mathcal{R}}_{testpoint}$ can be approximated using an appropriate inclusion f_{expand} of these points. As we will explain later, an inclusion operation is also needed while expanding the target regions to the actual state space regions (see Figure 6.14). These two steps can be combined. Even a few test points may give a reasonable target approximation, but the Region $\widetilde{\mathcal{R}}_{testpoint}$ might be under- or overestimated

$$\mathcal{P}_{test} = \{\mathbf{p}_1, \mathbf{p}_2, \ldots, \mathbf{p}_{n_t}\}, \ \mathbf{p}_i \in \mathcal{R}_{test}, \tag{6.40}$$

$$\mathcal{P}_{target} = \{s(\mathbf{p}_i) \mid \mathbf{p}_i \in \mathcal{P}_{test}\}, \tag{6.41}$$

$$\widetilde{\mathcal{R}}_{testpoint} = \{f_{expand}(\mathbf{x}^z) \mid \mathbf{x}^z \in \mathcal{P}_{ziel}\} \simeq \mathcal{R}_{exact}. \tag{6.42}$$

There are two approaches making this process rigorous which means that the target Region $\mathcal{R}_{exact}$ is fully included in the target approximation. At first, it is shown in [25] that $\widetilde{\mathcal{R}}_{testpoint}$ is certainly overestimated if all corner values are used as test points and if s can be assumed to be monotonic. Secondly, following the argumentation in [8], this is done using Lipschitz constants L in each state space dimension. Using a grid of test points, spaced by $\mathbf{h}$, we can calculate an extension diameter $\mathbf{d}_{ex} = \mathbf{k}_L \mathbf{h}$ for the target points. Expanding each target point by $\mathbf{d}_{ex}$ gives a set of boxes. The union of these boxes is an overestimated target approximation $\widetilde{\mathcal{R}}_{lipschitz}$. In Figure 6.13 three test and target points and the dedicated extension boxes are shown

$$\begin{aligned}\widetilde{\mathcal{R}}_{lipschitz} &= \{f_{expand}(s(\mathbf{x}^z), \mathbf{k}_L, \mathbf{h}) \mid \mathbf{x}^z \in f_{grid}(\mathbf{h}, \mathcal{R}_{test})\} \\ &\supseteq \mathcal{R}_{exact}.\end{aligned} \tag{6.43}$$

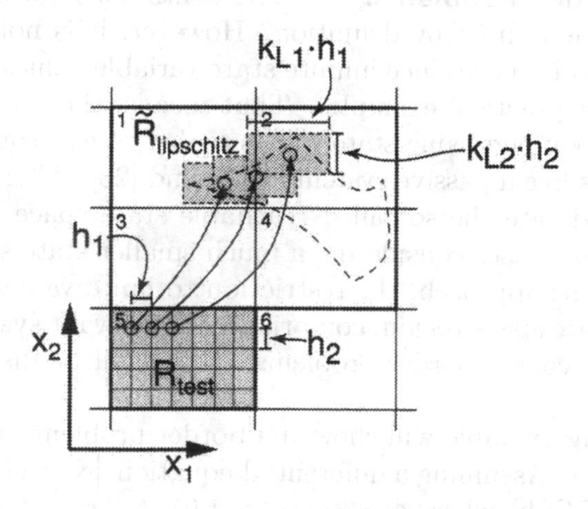

Figure 6.13. Optimized test point method using Lipschitz constants.

All discussed target Regions ($\mathcal{R}_{exact}$, $\widetilde{\mathcal{R}}_{testpoint}$, and $\widetilde{\mathcal{R}}_{lipschitz}$) do not match the state space subdivisions used. Therefore, a second step is necessary to extend these regions to a legal set of boxes. For example, the expansion of Region $\mathcal{R}_{exact}$ (hatched areas in Figure 6.14) is given by the set of all boxes having contact with the target Region $\mathcal{R}_{exact}$. Fortunately, this operation is always an overestimation and therefore does not impact the correctness of the above results. Up to now, only the approximation $\widetilde{\mathcal{R}}_{testpoint}$ has been used in the experimental results.

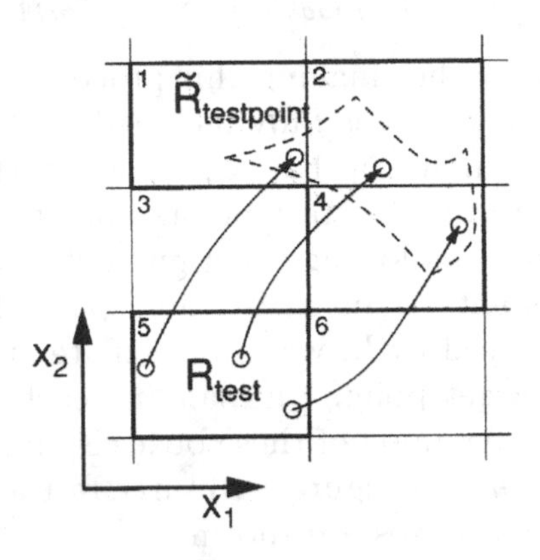

Figure 6.14. Expanded successor region matching the state space subdivisions.

4.2.5 Border Problems. The continuous state space for an analog system is infinite by definition. However, it is not necessary to consider the whole space since infinite state variable values are not physically useful for practical examples. That means, the system dynamics will always force the growing state values to decrease at some point. The system behaves like a passive system at the end [28]. There are different methods to estimate the so called reachable state space regions [7, 3]. However, in most cases considering a much smaller state space region is sufficient. In this approach, the restriction to finite values is done by a user defined state space region, comprising the relevant system behavior. The restriction causes border problems, which will be discussed in this section.

The following example will show the border problems and the implemented solution. Assuming a differential equation $\{\dot{\mathbf{x}} = (1,0)^T\}$ and the CTL formula $\mathsf{EG}(\Phi)$ where $\Phi = ((x_{(2)} > 1.0) \ \wedge \ (x_{(2)} < 2.0))$. For this simple example the model checking problem is easy to solve analytically. For a given time step δt the system solution $\mathbf{x}(t_n + \delta t)$ is determined by Equation (6.44)

$$\mathbf{x}(t_n + \delta t) = \mathbf{x}(t_n) + \int_{t_n}^{t_n+\delta t} \dot{\mathbf{x}}(\tau)\, d\tau = \begin{bmatrix} x_{(1)}(t_n) + \delta t \\ x_{(2)}(t_n) \end{bmatrix}. \quad (6.44)$$

It is obvious, that $\mathsf{EX}(\Phi) = \Phi$ for all δt, since a time step causes only a shift in $x_{(1)}$ direction and the Region Φ is not restricted there. $\mathsf{EG}(\Phi)$ is the largest fix point of the sequence $\{\Phi_0 = \Phi; \Phi_{i+1} = \Phi_i \wedge \mathsf{EX}(\Phi_i)\}$. Consequently, the analytical result of $\mathsf{EG}(\Phi)$ is Φ.

If a restricted state space - for example $([-5 \mathinner{..} 5], [-5 \mathinner{..} 5])^T$ - is applied to this example the result changes dramatically. Using the solution of $\mathbf{x}(t_n + \delta t)$ we find $\mathsf{EX}(\Phi)$ to be $([-5 \mathinner{..} (5 - \delta t)], [1 \mathinner{..} 2])^T$. Thus, the largest fix point for the sequence defining EG is $\varnothing$ (see Figure 6.15 left).

This result should be reviewed in detail since the state space restriction should not have an impact on the model checking result. Therefore, the meaning of $\Psi = \mathsf{EG}(\Phi)$ has to be studied again: "For each state ψ within Ψ there is a path starting in ψ such that Ψ is invariantly true on this path." In the given example, every path leaves the restricted state space after some time but that does not necessarily mean that Ψ is not fulfilled on that path since Ψ is also restricted to the given state space. Thus, it is a free choice to define whether a path leaving or entering the restricted state space region fulfills a given CTL condition or not.

In other words, it has to be defined whether the area outside of the restricted state space is part of the actual region being the argument of the CTL-A function or not. In the above example we simply assumed that the outside area is not part of the Region Φ

$$\begin{aligned} \Phi &= \left(([-5 \mathinner{..} 5], [1 \mathinner{..} 2])^T \cap \mathit{Outside} \right), \\ \mathsf{EG}(\Phi) &= \varnothing. \end{aligned} \tag{6.45}$$

Under this assumption the given result is correct (Equation (6.45)). If we assume the outside area to be part of the Region Φ we get the result shown in Equation (6.46) (see Figure 6.15 right)

$$\begin{aligned} \Phi &= \left(([-5 \mathinner{..} 5], [1 \mathinner{..} 2])^T \cup \mathit{Outside} \right), \\ \mathsf{EG}(\Phi) &= \Phi. \end{aligned} \tag{6.46}$$

To implement this function, an outside area flag is stored for each region used during the CTL-A evaluation. Every newly defined region will by default exclude the outside area and the outside area flag is false in this case. All boolean operations are not only applied to the regions but also to the outside area flags. Thereby, it is possible to define all possible constellations of regions and outside area flags.

4.2.6 Input Value Model. To solve Equation (6.35) or (6.36) the input value $\mathbf{u}(t)$ is needed. Until now, we have not defined this value.

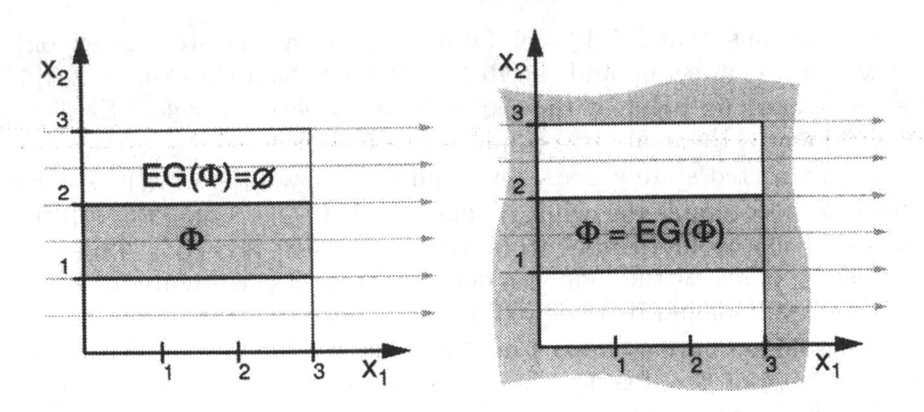

Figure 6.15. Border problem example in a restricted state space. Left: The outside area is not part of Φ. Right: The outside area is included in Φ.

In principle, the input signals might be defined explicitly. However, this is not really useful since the model checking result will only be true for one specific input signal and this is a contradiction to the formal verification idea. Therefore, some conditions for the input signals are assumed without defining them explicitly. To do this, the state space is extended by the input variables resulting in the extended state space (see Section 2.2). Thus, every state within the extended state space contains information on the actual input values. However, there is no information on the input value change with respect to time. Moreover, it is theoretically impossible to predict the input value variation because the input values are not determined by the system itself but rather from some outside systems.

There are two extreme assumptions: The input values do not change at all and the input values may change over the whole input value range. For the first assumption, the model is built up for several constant input values as described before. There will not be any transition between states with different input values. In the second approach, a state space region has not only transitions to regions at the same input level but additional transitions to the neighbor regions in terms of input values. By using the extended state space and the described input model the transition relation changes to $R \subseteq \left(\mathbb{R}^{n+i} \times \mathbb{R}^1 \times \mathbb{R}^{n+i}\right)$.

As we will see later, both of these input models are useful for certain conditions to be checked. Between these two extreme models it is possible to assume the input values to vary within a given frequency range or within a maximum input voltage slope.

4.2.7 Optimizations. The algorithms described above, generates a successor operation for the discrete state space model. However, experiences show that some additional steps are necessary to optimize the transition relation for some corner cases. Namely, these are prevention of long successor vectors, resulting in a box over-jump, boxes with self-connection, and boxes with no transitions to other boxes.

Due to long successor vectors in terms of box diameters a neighbor box over-jump can occur. The results are unwanted holes within the CTL output regions. Without losing generality, the successor vector length can be reduced preventing this behavior.

As we have seen before, the subdivision algorithms store information about boxes possibly containing fix points (Section 4.2.3). For all remaining boxes (not containing fix points) it is unphysical to have a self-connection or no connection to other neighbor boxes. Thus, all self-connections within these boxes are removed. By applying additional subdivisions in regions without connections an optimized transition relation is generated.

It has been mentioned in Section 4.2.2, that no explicit time relations are considered. It might be useful and necessary in future to store not only the transition relation $R \subseteq \mathcal{Q} \times \mathcal{Q}$ but rather this relation combined with the related transition time delays $R \subseteq \mathcal{Q} \times Z \times \mathcal{Q}$ where Z denotes the set of all transition time delays used.

4.3 Experimental Results

Two small nonlinear examples are used to show the capability of the proposed algorithms. The first one is a Schmitt trigger circuit and the second one is a tunnel diode oscillator.

4.3.1 Schmitt Trigger Example. The circuit used is described in Section 3.4.1. We use the netlist representation for model checking. To consider all states that might occur, the extended state space is chosen to be $V_{out} = [-7.7 .. 7.7]$ and $V_{in} = [-7.7 .. 7.7]$. The most interesting features of the Schmitt trigger function are the switching properties for one output state to the other. The result of formulating this by CTL-A is: $\Phi_1 = \mathsf{EF}(V_{out} < -4.5)$. Φ_1 is the set of states in which a path exists that will eventually reach the region $V_{out} < -4.5$. We choose the constant input value model for this calculation. The collection of boxes fulfilling this condition is shown in Figure 6.16 in light gray.

Obviously, the circuit always switches to the negative output state above $V_{in} \simeq 2V$. Below this point the switching depends on the output state V_{out}. This behavior can directly be extracted from Figure 6.6

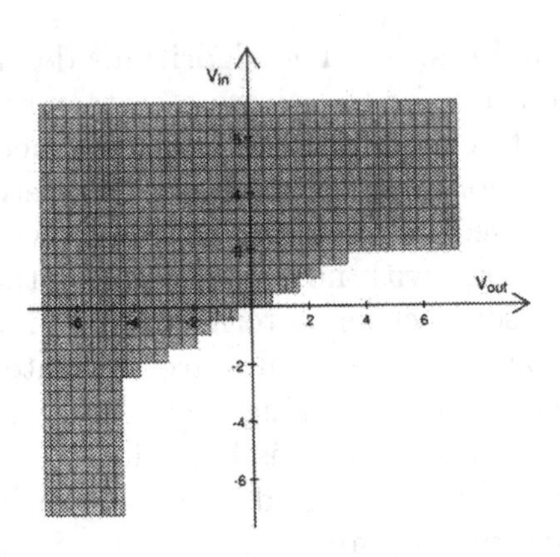

Figure 6.16. Model checking result Φ_1.

showing the unstable equilibriums in the circuit. An equivalent formula can be applied to find the positive switching conditions.

The result of a CTL-A formula is the collection of boxes fulfilling this condition. Such graphical output might be useful for analog designers. In general however, we would expect only *true* or *false* as output for a CTL-A formula. To derive this, the checking condition can be expanded by an additional statement. For example

$$\Phi_2 = \mathsf{EF}(V_{out} < -4.5)\ \&\ ((V_{out} > -1)\ \&\ (V_{in} < -1)). \tag{6.47}$$

In this case, the output is an empty set, which means there is no path from region $((V_{out} > -1)\ \&\ (V_{in} < -1))$ to the negative output state. CTL-A formula (6.47) is *false* for the whole state space proving that the Schmitt trigger circuit has a stable output for the given input region. If the output set is not empty then at least one state fulfills the formula. Every CTL-A formula can be changed in the same manner to obtain a binary result instead of a graphical one. However, we prefer graphical results since they can give more insight into the algorithms.

4.3.2 Tunnel Diode Oscillator Example. The analog system in our second example is a simple tunnel diode oscillator circuit shown in Figure 6.17. The input voltage V_{in} is set to 2.6 V. In this operating point the circuit starts to oscillate automatically. The bounded state space is given by $V_C = [-0.2 \,..\, 4.4]$ and $I_L = [-0.2 \,..\, 4.0]$.

According to digital systems a stable oscillation might be proved by the following CTL-A equation

$$\Phi_3 = \{\mathsf{AG}(\mathsf{AF}(I_L > 2.2))\ \&\ \mathsf{AG}(\mathsf{AF}(I_L < 1.6))\}\,. \tag{6.48}$$

The collection of boxes fulfilling this condition is shown in the left part of Figure 6.18 in light gray. Except of some border boxes and the middle region, the whole state space is covered. The border boxes cannot be considered in the verification result due to the algorithm described in Section 4.2.5. The empty middle region is caused by the limited resolution of the discrete model. The algorithm detects that fix points are possibly included in these state space regions. Therefore, the above CTL-A formula is not fulfilled. Theoretically, only one single fix point is present in this region.

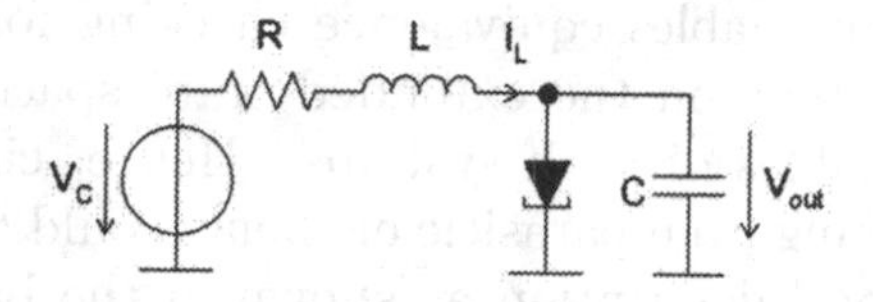

Figure 6.17. **Tunnel diode oscillator.**

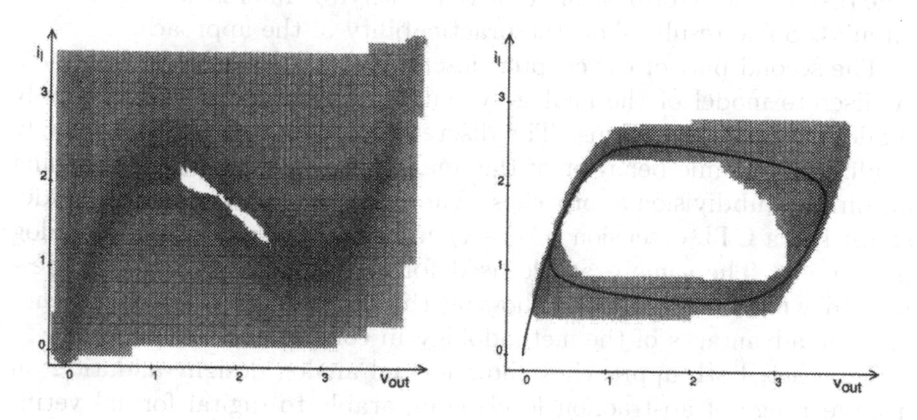

Figure 6.18. Left: Model checking result Φ_3. Right: Model checking result Φ_4.

We can conclude that nearly the whole plane will float into an stable orbit. The next question might concern the possible orbit geometry. We generate this by applying $\Phi_4 = \left\{(\mathsf{EG}^{-1}(\Phi_3))\right\}$. The result Φ_4 contains the whole orbit calculated by an ordinary simulation (black line in Figure 6.18 right).

These two examples illustrate possible property descriptions. It seems to be possible to describe a lot of properties in the time domain like large signal output swing, PSRR (power supply rejection ratio), stability, oszillations etc. with CTL-A. Properties with numeric timing informations like slew rate need the mentioned extension of our algorithm (see Section 4.2.7). Until now, frequency domain properties are not considered. These topics are future research work.

5. Summary

In this chapter, two approaches to formal verification of analog systems with nonlinear dynamic behavior are presented. Both approaches can deal with strongly nonlinear systems like Schmitt trigger circuits, which is shown in the example sections. The abstraction level of the handled systems ranges from transistor netlists to behavioral descriptions in a hardware description language.

The first approach enables equivalence checking for analog systems. It compares two systems in the extended state space spanned by the dominant state variables of both systems. Hence, circuits with many state variables resulting from parasitic elements could be verified against an abstract behavioral description as shown in the bandpass example. The equivalence checking procedure samples the state space. An explicit conversion into a discrete model for verification purposes is not necessary. The result is an error measure clearly identifying differences between the circuits. Some results show the practicability of the approach.

The second part of the chapter describes a model checking approach. A discrete model of the analog system is generated in order to apply model checking algorithms. The discrete modeling retains the strongly nonlinear dynamic behavior of the analog circuit with few states using automatic subdivision approaches. The properties to be checked are described in a CTL extension (CTL-A) enabling the description of analog properties. The same example used for equivalence checking is investigated with model checking showing the differences of the approaches and the advantages of the methodology in comparison to simulation.

Together, both approaches allow formal analog design validation on a large range of abstraction levels comparable to digital formal verification methodology. Furthermore, an integration with digital formal verification is possible. Accordingly, an implementation of mixed-signal algorithms based on these approaches seems to be the next step towards a closed formal verification flow.

6. Acknowledgement

The authors would like to thank Dr. Uwe Feldmann, Infineon Technologies, and his team for fruitful discussions and contributions on efficient state space calculation and system transformation to a Kronecker Canonical Form.

Appendix: Mathematical Symbols

$\mathbb{R}$	Set of real numbers
$\mathbf{x}$	Vector
$\mathbf{x}_{(i)}$	ith component of vector $\mathbf{x}$
$\mathbf{A}$	Matrix
$f(x)$	Function in terms of x
$\mathbf{f}(x)$	Vector of functions
$\dot{x} = \frac{dx(t)}{dt}$	Time derivative of $x(t)$
$\mathcal{A}$	Set
A	CTL operator
Φ	Temporal logic expressions
$[x_l..x_u]$	Interval with lower bound x_l and upper bound x_u

Special variable naming

$\mathbf{x}^{\mathsf{z}}$	Vector of state variables
$\mathbf{x}^{\mathsf{a}}$	Vector of algebraic variables
$\mathbf{z}$	Vector of virtual state variables

References

[1] R. Alur, T.A. Henzinger, G. Lafferriere, and G.J. Pappas. Discrete abstractions of hybrid systems. *Proceedings of IEEE*, (88):971–984, 2000.

[2] K.J. Antreich, H.E. Graeb, and C.U. Wieser. Circuit analysis and optimization driven by worst-case distances. *IEEE Transactions on Computer-Aided Design of Integrated Circuits and Systems*, 13(1):57–71, 1994.

[3] E. Asarin, O. Bournez, T. Dang, and O. Maler. Approximate reachability analysis of piecewise-linear dynamical systems. *HSCC '00: Hybrid Systems: Computation and Control, LNCS*, pages 76–90, 2000.

[4] P.N. Brown, A.C. Hindmarsh, and R.P. Linda. Consistent initial condition calculation for differential-algebraic systems. *SIAM Journal on Scientific Computing*, 19(5):1495–1512, 1998.

[5] F.H. Bursal and B.H. Tongue. A new method of nonlinear system identification using interpolated cell mapping. *ACC '92: American Control Conference*, 4:3160–3164, 1992.

[6] E.M. Clarke and E.A. Emerson. Design and synthesis of synchronisation skeletons using branching time temporal logic. *Lecture Notes in Computer Science, Springer-Verlag*, 131, 1981.

[7] T. Dang and O. Maler. Reachability analysis via face lifting. *HSCC '98: Hybrid Systems: Computation and Control, Lecture Notes in Computer Science*, pages 96–109, 1998.

[8] M. Dellnitz, G. Froyland, and O. Junge. The algorithms behind gaio - set oriented numerical methods for dynamical systems. *Ergodic Theory, Analysis, and Efficient Simulation of Dynamical Systems (eds. B. Fiedler), Springer*, pages 145–174, 2001.

[9] A. Dharchoudhury and S.M. Kang. Worst-case analysis and optimization of VLSI circuit performances. *IEEE Transactions on Computer-Aided Design of Integrated Circuits and Systems*, 14(4):481–492, 1995.

[10] P. Van. Dooren. The computation of Kronecker's canonical form of a singular pencil. *Journal on Linear Algebra and its Applications*, 27:103–140, 1979.

[11] D. Estévez-Schwarz. Consistent initialization for index-2 differential algebraic equations and its application to circuit simulation. *Dissertation, Humboldt-Universität Berlin*, 2000.

[12] P. Feldmann and R.W. Freund. Efficient linear circuits analysis by pade approximation via the lanczos process. *IEEE Transactions on Computer-Aided Design of Integrated Circuits and Systems*, 14(5):639–649, 1995.

[13] L. Fortuna, G. Nunnari, and A. Gallo. Model order reduction techniques with applications in electrical engineering. *Springer-Verlag, Berlin*, 1992.

[14] C.W. Gear. Differential algebraic equations, indices, and integral algebraic equations. *SIAM Journal on Numerical Analysis*, 27(6):1527–1534, 1990.

[15] M. Günther and U. Feldmann. The DAE-index in electric circuit simulation. *Mathematisches Institut, Technische Universität München*, 1993.

[16] W. Hartong, L. Hedrich, and E. Barke. Model checking algorithms for analog verification. *DAC '02: Design Automation Conference*, pages 542–547, 2002.

[17] W. Hartong, L. Hedrich, and E. Barke. On discrete modeling and model checking for nonlinear analog systems. *CAV '02: International Conference on Computer-Aided Verification, LNCS*, 2404:401–413, 2002.

[18] L. Hedrich and E. Barke. A formal approach to nonlinear analog circuit verification. *ICCAD '95: International Conference on Computer Aided Design*, pages 123–127, 1995.

[19] L. Hedrich and E. Barke. A formal approach to verification of linear analog circuits with parameter tolerances. *DATE '98: Design, Automation and Test in Europe*, pages 649–654, 1998.

[20] L. Hedrich and W. Hartong. Approaches to formal verification of analog circuits. *Low-Power Design Techniques and CAD Tools for Analog and RF Intergrated Circuits, Wambacq, P., eds., Kluwer Academic Publishers, Boston*, 2001.

[21] T.A. Henzinger and P.-H. Ho. Algorithmic analysis of nonlinear hybrid systems. *CAV '95: International Conference on Computer-Aided Verification, LNCS*, 939(7):225–238, 1995.

[22] T.A. Henzinger, P-H. Ho, and H. Wong-Toi. Hytech: A model checker for hybrid systems. *Lecture Notes in Computer Science, Springer-Verlag*, pages 460–463, 1997.

[23] C.W. Ho, A.E. Ruehli, and P.A. Brennan. The modified nodal approach to network analysis. *IEEE Transactions on Circuits and Systems*, 22(6):504–509, 1975.

[24] T. Kropf. Introduction to formal hardware verification. *Springer-Verlag, Berlin, Heidelberg*, 1999.

[25] R.P. Kurshan and K.L. McMillan. Analysis of digital circuits through symbolic reduction. *IEEE Transactions on Computer-Aided Design of Integrated Circuits and Systems*, 10(11):1356–1371, 1991.

[26] F. Laroussinie and P. Schnoebelen. Specification in CTL+past for verification in CTL. *Information and Computation*, 156(1/2):236–263, 2000.

[27] R. März. Numerical methods for differential algebraic equations. *Acta Numerica*, pages 141–198, 1991.

[28] W. Mathis. Theorie nichtlinearer Netzwerke. *Springer-Verlag, Berlin*, 1987. (German).

[29] K.L. McMillian. Symbolic model checking. *Kluwer Academic Publishers, Boston*, 1993.

[30] S. Natarajan. A systematic method for obtaining state equations using MNA. *IEE Proceedings G*, 138(3):341–346, 1991.

[31] L. Petzold. Differential/algebraic equations are not ODE's. *JSIAM*, 3(3):367–384, 1982.

[32] A. Puri and P. Varaiya. Decidability of hybrid systems with rectangular differential inclusions. *CAV '94: International Conference on Computer-Aided Verification, LNCS*, pages 95–104, 1994.

Index